FORESTRY COMMISSION BULLETIN 95

Forest Fertilisation in Britain

C. M. A. Taylor
Silviculturist,
Forestry Commission

LONDON: HMSO

© *Crown copyright 1991*
  First published 1991

ISBN 0 11 710294 6
ODC 237.4 : 114.54 : 424.7 : (410)

KEYWORDS: Nutrients, Forestry

Enquiries relating to this publication
should be addressed to:
The Technical Publications Officer,
Forestry Commission, Forest Research Station,
Alice Holt Lodge, Wrecclesham,
Farnham, Surrey GU10 4LH.

Front cover: Aerial application of fertiliser to recently planted crop. *(C. P. Quine)*
Inset. Good growth response in Sitka spruce to an application of phosphate. *(J. M. Mackenzie)*

Contents

	Page
Summary	v
Résumé	vi
Zusammenfassung	viii
1 **Introduction**	1
2 **Recognition of nutrient deficiency**	3
Phosphorus	3
Potassium	3
Nitrogen	4
Other nutrients	6
Foliar nutrient levels	7
3 **When to fertilise**	8
4 **Where to apply fertiliser**	10
Scotland and northern England	10
Phosphate and potassium fertiliser	11
Grass-dominated mineral soils	11
Heathland mineral soils	13
Shallow peats	13
Deep peats	14
Nitrogen fertiliser	14
Categorisation of nitrogen-deficient sites	15
Correct choice of species and nursing mixtures	17
Nitrogen-fixing plants	18
Sewage sludge	19
Application of lime	20
Wales	20
Southern England	20
Restocking	21
Unusual sites	22
Man-made soils	22
Littoral soils	22
High elevations	22

5 **Types and rates of fertiliser**	23
Phosphorus	23
Soluble vs. insoluble	23
Rate and season of application	24
Potassium	24
Types	24
Rate and season of application	24
Nitrogen	24
Urea vs. ammonium nitrate	24
Rate and season of application	25
Other nitrogen sources	25
Compound fertilisers	25
6 **Methods of application**	27
Aerial application	27
Application from the ground	28
7 **Economic appraisal**	30
The 'years saved' method	30
Justifying fertiliser costs	30
Permanent change in Yield Class	31
8 **Environmental effects**	32
9 **Effects on wood quality**	33
Acknowledgements	34
References	35
Appendix I - Foliar sampling procedure	40
Appendix II - Forestry Commission foliar analysis service	42
Appendix III - Forestry Commission site record form for use with shoot samples for foliar analysis	43

Forest Fertilisation in Britain

Summary

In Britain the use of fertilisers has greatly increased the productivity of forests growing on nutrient-poor soils. In fact, many sites could not otherwise have been successfully afforested. From the early pioneering work of Stirling-Maxwell to the present day, the Forestry Commission has continually tested rates and types of fertiliser and methods of application. A pattern has gradually emerged from these empirical experiments indicating the fertiliser requirements of the main tree species planted. This has been aided by complementary basic research on forest nutrition, particularly at the Macaulay Institute, Edinburgh University and the Institute of Terrestrial Ecology. This Bulletin attempts to condense this research into practical guidance for the forest manager.

The symptoms of the most common nutrient deficiencies (nitrogen, phosphorus and potassium) are described as well as the methods of treatment. Fertiliser regimes for the application of phosphate and potassium are detailed by geographical region, soil type, lithology, age and species of trees. The treatment of nitrogen deficiency is indicated by the same parameters and further guidance is given on the choice between remedial herbicide treatment on heather or fertiliser application. The principal fertilisers are listed, together with the appropriate rate, season and method of application.

Other methods of relieving or preventing nutrient deficiency are described. These include the correct choice of tree species, mixtures of different species and the use of organic fertilisers. Some of these alternative treatments are well proven while others are still at the experimental stage. However, they do offer lower input systems on infertile sites while maintaining acceptable levels of timber production.

The main options for the economic appraisal of fertilising are explained, as well as the likely effects on the environment and wood quality. It is vital that foresters maintain rigorous standards when applying fertiliser to prevent wastage or pollution of watercourses.

This Bulletin is intended to supplement field experience and to aid rational decision-making. It is designed to present current knowledge in a structured fashion to assist programme planning at both regional and local level. It can also be used for indicating nutrient requirements on specific sites, although interpretation will require greater care and field verification will be essential.

Fertilisation des Fôrets en Grande Bretagne

Résumé

En Grande Bretagne on a augmenté considérablement la productivité des fôrets établies sur des sols pauvres en substances nutritives grâce à l'emploi des engrais. En fait, sans ceci, il y a plusieurs endroits où on n'aurait pu reboiser. Depuis le travail entrepris par le pionnier Stirling-Maxwell jusqu'au travail accompli aujourd'hui, la Forestry Commission analyse continuellement les quantités et les genres d'engrais aussi bien que les moyens dont ils sont utilisés. Suivant les résultats de ces essais empiriques, on a vu sortir petit à petit des caractéristiques qui ont indiqué la nature des engrais dont ont besoin les espèces principales d'arbres plantés. Ces découvertes ont été renforcées par des recherches complémentaires fondamentales sur la nutrition des forêts faites en particulier au Macaulay Institute, Université d'Edimbourg et à l'Institute of Terrestrial Ecology. Ce bulletin a comme but de tenter une synthèse de ces recherches pour en créer des conseils pratiques pour les gérants des forêts.

On y décrit les symptômes du manque des plus connues des substances nutritives (l'azote, le phosphore et le potassium) aussi bien que les moyens de le traiter. Des régimes d'engrais en ce qui concerne l'introduction du phosphate et du potassium sont indiqués en détail selon la région géographique, les caractéristiques du sol, la lithologie et l'âge et l'espèce des arbres. Le traitement pour le manque d'azote est indiqué selon les mêmes critères et on offre des conseils supplémentaires sur le choix entre un traitement curatif d'herbicide pour la bruyère et l'application des engrais. On donne une liste des engrais principaux ainsi que les quantités respectives, la saison et les moyens d'application.

On y décrit également d'autres moyens de remédier ou d'éviter le manque de substances nutritives. Ceci comprend le choix correct de l'espèce d'arbre, du mélange de plusieurs espèces et de l'emploi des engrais organiques. Certains d'entre ces traitements alternatifs se sont déjà avéré efficaces tandis que d'autres sont toujours au stade expérimental. Ils offrent néanmoins des systèmes qui réduisent les quantités utilisées sur les endroits infertiles tout en maintenant des niveaux acceptables de rendement du bois.

On explique les choix principaux pour l'évaluation économique de la fertilisation aussi bien que les effets probables sur l'environnement et la qualité du bois. Il est indispensable que les forestiers maintiennent des niveaux rigoureux quand il s'agit d'employer les engrais afin d'empêcher le gaspillage ou la pollution des cours d'eau.

Ce Bulletin est destiné à ajouter à l'expérience acquise sur le terrain et à aider à prendre de bonnes décisions. Il est conçu de façon à fournir les connaissances actuelles sous une formule structurée qui aidera l'organisation des programmes aux niveaux régionaux et locaux. Il peut servir également à indiquer les besoins en substances nutritives aux endroits déterminés bien qu'il faille prêter plus d'attention à l'interprétation et que la vérification sur terrain soit indispensable.

Walddüngung in Großbritannien

Zusammenfassung

In Großbritannien hat der Gebrauch von Düngemitteln die Produktivität der auf nährstoffarmen Böden wachsenden Wälder stark verbessert. Manche Gelände hätten sonst gar nicht erfolgreich aufgeforstet werden können. Von den frühen Pionierarbeiten Stirling-Maxwells bis zum heutigen Tag hat die Forstverwaltung ständig die Leistung und Arten von Düngemitteln und Anwendemethoden geprüft. Aus diesen empirischen Experimenten hat sich allmählich ein Muster entwickelt, das die Düngemittelerfordernisse der hauptsächlich angepflanzten Baumarten anzeigt. Das wurde besonders durch eine ergänzende Basisforschung über Waldnährstoffe im Macaulay Institute, Edinburgh University und dem Institute of Terrestrial Ecology unterstützt. Dieser Bericht versucht, diese Forschungsergebnisse in praktische Anweisungen für den Forstleiter zusammenzufassen.

Die Symptome der meisten Nährstoffmängel (Stickstoff, Phosphor und Kalium) sowie ihre Behandlungsmethoden werden hier beschrieben. Die Leistungsbereiche der Düngemittel beim Gebrauch von Phosphat und Kalium werden geographischen Gebieten, Bodenarten, Gesteinsarten sowie Baumalter und -arten entsprechend dargelegt. Die Behandlung von Stickstoffmangel wird durch dieselben Parameter angezeigt, und weitere Hilfe wird bei der Wahl zwischen abhilfsmäßiger Unkrautbekämpfung von Heide, und Düngemittelanwendung gegeben. Die hauptsächlich verwendeten Düngemittel werden aufgeführt sowie die entsprechende Leistung, Jahreszeit und Anwendungsmethode.

Weitere Methoden zur Minderung oder Verhinderung von Nährstoffmangel werden beschrieben. Diese sind u.a. die richtige Wahl der Baumarten, die Mischung der verschiedenen Arten und der Gebrauch organischer Düngemittel. Einige dieser alternativen Behandlungsmethoden haben sich gut bewährt und andere befinden sich noch im Versuchsstadium. Sie bieten zwar niedrigere Eingabesysteme auf unfruchtbaren Geländen, erhalten aber ein akzeptables Niveau bei der Holzproduktion.

Die hauptsächlichen Möglichkeiten für die wirtschaftliche Beurteilung von Düngung werden erklärt sowie ihre wahrscheinlichen Auswirkungen auf die Umwelt und Holzqualität. Es ist außerordentlich wichtig, daß Förster bei der Verwendung von Düngemitteln strenge Maßstäbe wahren, um einen Verlust oder eine Verschmutzung der Wasserwege zu verhindern.

Dieser Bericht ist dafür gedacht, die Erfahrung im Feld zu unterstützen und eine rationale Entschlußfassung zu unterstützen. Er ist so konstruiert, daß er die gegenwärtige Kenntnis auf eine strukturierte Weise darlegt, um der Programmplanung auf regionalem und lokalem Niveau zu helfen. Er kann auch zur Anzeige von Nährstofferfordernissen auf bestimmten Geländen verwendet werden, obwohl die Interpretation eine größere Genauigkeit benötigt, und ein Nachweis im Feld notwendig sein wird.

Chapter 1
Introduction

There has been a considerable afforestation programme in Great Britain since the introduction of the Forestry Act in 1919. The area of productive woodland now stands at over 2 million hectares, of which 55% is privately owned and 45% is state owned and managed by the Forestry Commission. The early planting was mainly on fertile soils at lower elevations where good growth rates were achieved with little or no fertiliser input. After the 1950s, improved establishment techniques enabled an expansion on to poorer quality land, particularly peatland. However, it soon became obvious that application of fertiliser would be required on these sites to ensure satisfactory rates of growth.

The need for additional nutrients on certain afforestation sites was recognised before the First World War when Sir John Stirling-Maxwell experimented with basic slag (Stirling-Maxwell, 1925). Trials of nitrogen and potassium by the Forestry Commission in the 1930s were abandoned after many trees were killed, and research was concentrated on phosphate fertilisers. Basic slag, triple superphosphate, superphosphate and ground mineral phosphate were all tested at various rates and methods of application (Zehetmayr, 1960). In the period from 1945 to 1949, investigations by Dr E. M. Crowther of Rothamsted on the nutrient requirements of nursery stock were carried over into the forest in a series of 'nursery extension' experiments. Crowther re-examined application techniques and found that placement of fertiliser did not yield any advantage over broadcasting. He also tested compound fertilisers containing nitrogen, phosphate and potassium and demonstrated that the main benefit was from the application of phosphate.

In the 1960s the Forestry Commission Research Division and the Macaulay Institute for Soil Research (now Macaulay Land Use Research Institute) showed the beneficial effect of application of potassium on deep peats, and investigated nitrogen deficiency in certain tree genera, particularly spruce (*Picea* spp.), on unflushed peats and heathland mineral soils. As yet, deficiency of other nutrients has either not been recorded or has only occurred as a small, localised problem (Binns *et al.*, 1980), with the exception of copper deficiency which has been reported on several occasions (Taylor, 1989).

Application of fertilisers is now regarded as a standard treatment when establishing trees on infertile soils in Great Britain (Plates 1 and 2 show what can happen to unfertilised crops on such sites). Currently, over 50 000 ha are treated each year (Table 1), the major application being phosphate. The large areas treated with potassium fertiliser indicate the high proportion of planting that has taken place on peatland in the 1970s and 1980s. Fertilising with nitrogen is extremely important on certain site types and there is evidence that this programme should be increased (Taylor, 1990a).

Table 1. Forest fertiliser programme in Great Britain (1981-86)

Element applied	Average annual area fertilised (ha)
Phosphorus	25 000
Phosphorus and potassium	29 000
Nitrogen	2 000
Total	**56 000**

Plate 1. Sitka spruce stand (25 years old) suffering severe growth check due to nitrogen deficiency. (P. M. Tabbush)

Plate 2. Sitka spruce stand (55 years old) suffering severe growth check due to phosphorus deficiency. (C. M. A. Taylor)

Chapter 2
Recognition of Nutrient Deficiency

The ability to recognise nutrient deficiency visually is an essential skill for all forest managers. It will aid decisions on the timing and type of fertiliser application required. Foliar analysis is useful for confirmation of visual assessment in complex or abnormal cases. It is important to prevent severe nutrient deficiency in crops as these will take longer to respond to fertiliser application, particularly if foliage has been lost.

Phosphorus

The principal functions of phosphorus are in energy transfer and as a constituent of nucleic acids (Baule and Fricker, 1970; Bould et al., 1983). It is not very mobile within the tree and thus is not readily transferred from older tissue to the new shoots (Binns et al., 1980).

Phosphorus deficiency does not appear as a discoloration, but as a reduction in the size of needles/leaves and in general growth of the tree. In severe cases coniferous trees become stunted with needles closely adpressed to the stem giving a scale-like appearance (Plate 3). Loss of older needles can also occur and, occasionally, there will be dieback of leading shoots (Plate 4). In severe deficiency in broadleaves purple or red tints may appear, particularly at leaf margins (Binns et al., 1983).

Potassium

Potassium is thought to be important for balancing charges within plant cells, control of stomata, protein synthesis and as an activator for several enzymes (Bould et al., 1983). It is a very mobile element within the tree and can be withdrawn from older needles and transferred to those on young extending shoots (Binns et al., 1980).

In young conifers potassium deficiency shows as a yellowing of current foliage extending from the tip of the needles. In severe cases the whole shoot can be affected. Loss of apical dominance (i.e. no distinctive leading shoot) is also common at this stage (Plate 5). When the trees are older the yellowing is confined to the older foliage and the current foliage displays normal growth and colour (Plate 6). Potassium defi-

Plate 3. *Phosphorus deficient Sitka spruce branch.* (C. M. A. Taylor)

Plate 4. *Shoot dieback and loss of needles on Sitka spruce caused by severe phosphorus deficiency.* (R. McIntosh)

ciency has a lesser effect on height growth than either nitrogen or phosphorus deficiency but diameter growth can be severely reduced. In cases of severe potassium deficiency in thicket stage crops considerable needle loss and shoot death of the older foliage can occur (Plate 7). Tree crowns become very thin and interception of rainfall is much reduced, leading to raised water tables in high rainfall areas. Rootable volume and availability of potassium are then reduced and crops suffer a severe decline in growth.

In broadleaves potassium-deficient leaves are yellowish, sometimes between the veins, with marginal scorch in severe cases. Blue or purple tints may also appear. Protected leaves in the inner crown may be less affected than exposed leaves (Binns *et al.*, 1983).

Nitrogen

Nitrogen is a constituent of some of the most important substances within plants, including proteins, nucleic acids and chlorophyll (Baule and Fricker, 1970). It is an essential component of metabolic processes producing new tissue and consequent increase in wood volume.

Plate 5. *Loss of apical dominance in Sitka spruce due to severe potassium deficiency.* (C. M. A. Taylor)

Nitrogen in its various ionic forms is relatively mobile within plants so that trees are able to draw on stored reserves for transfer to the growing points (Bould et al., 1983).

The first indication of nitrogen deficiency is a lightening of the foliage colour from dark green to light green because of the effect on chlorophyll production. In severely deficient trees the foliage turns yellow-green or completely yellow. This colour change takes place over the whole tree and needle/leaf size and shoot extension are greatly reduced. These symptoms are typically developed in Sitka spruce (*Picea sitchensis* (Bong.) Carr) (Plate 8), other spruces, Douglas fir (*Pseudotsuga menziesii* (Mirb.) Franco) and other firs (*Abies* spp.) on sites low in available nitrogen. Symptoms of nitrogen deficiency are rarely seen in pines

Plate 6. *Potassium deficient Sitka spruce branch showing that yellowing is confined to older needles.* (C. M. A. Taylor)

Plate 7. *Loss of older needles and death of side shoots on a Sitka spruce branch caused by severe potassium deficiency.* (C. M. A. Taylor)

Table 2. Deficient and optimal foliar nutrient concentrations for conifers and broadleaves, as per cent oven-dry weight (from Binns et al., 1980 and 1983)

Species	Nitrogen Def.	Nitrogen Opt.	Phosphorus Def.	Phosphorus Opt.	Potassium Def.	Potassium Opt.
Sitka spruce / Norway spruce	<1.2	>1.5	<0.14	>0.18	<0.5	>0.7
Lodgepole pine / Scots pine	<1.1	>1.4	<0.12	>0.14	<0.3	>0.5
Corsican pine	<1.2	>1.5	<0.12	>0.16	<0.3	>0.5
*Douglas fir	<1.2	>1.5	<0.18	>0.22	<0.6	>0.8
*Western hemlock	<1.2	>1.5	<0.25	>0.30	<0.6	>0.8
*Japanese larch / *Hybrid larch	<1.8	>2.5	<0.18	>0.25	<0.5	>0.8
*Alder	<2.5	>2.8	<0.16	>0.18	<0.7	>0.9
*Birch	<2.5	>2.8	<0.19	>0.22	<0.7	>0.9
*Oak	<2.0	>2.3	<0.14	>0.16	<0.7	>0.9
*Ash / *Norway maple	<2.0	>2.3	<0.19	>0.22	<0.7	>0.9
*Beech / *Sweet chestnut	<2.0	>2.3	<0.14	>0.16	<0.7	>0.9

Notes: a. Values are tentative for the species marked*
b. Values between deficient and optimum levels are described as marginal.

Plate 8. *Nitrogen deficient Sitka spruce tree.* (C. M. A. Taylor)

(*Pinus* spp.) or larches (*Larix* spp.) and at worst occur as a temporary loss of colour and reduction in growth rate. They are also rare in broadleaved crops as these are normally planted on reasonably fertile sites.

Other nutrients

Deficiencies of the other major nutrient elements, i.e. magnesium, sulphur and calcium, are rarely found in forest crops although magnesium deficiency can be encountered on certain soils in southern England and in nurseries (Binns *et al.*, 1980; Benzian and Warren, 1956). Deficiencies of trace elements have not been recorded, except for copper on a few restocking sites, nurseries and flushed peat soils (Benzian, 1965; Binns *et al.*, 1980). However, recent investigations have revealed significant local copper deficiency in restocking sites on deep peats and peaty gleys over Carboniferous drift in the north-east of England (Taylor, 1989; Zulu, 1989). This is manifested as loss of apical dominance and the occurrence of contorted shoots (Plate 9).

Plate 9. Shoot distortion in Sitka spruce due to copper deficiency. (C. M. A. Taylor)

Foliar nutrient levels

The approximate deficient and optimal nutrient levels for conifers (Binns *et al.*, 1980) and broadleaves (Binns *et al.*, 1983) are presented in Table 2. When the foliar nutrient levels are above optimum, it is unlikely that the crop will respond to fertiliser. However, at marginal levels, deficiency symptoms will start to show and trees with deficiency levels will certainly benefit from application of the appropriate nutrient.

The levels for conifers were determined from samples taken from tree crops less than 5 m tall. There is now evidence that conventional foliar analysis techniques do not give useful indications of foliar nutrient status in older crops, particularly spruce (McIntosh, 1984a). By this stage nutrient cycling is fully developed and sampling from the upper crown is not only difficult but can result in overestimation of the nutrient status. McIntosh (1984a), however, did find good correlations between foliar nutrient levels in pole-stage Scots pine (*Pinus sylvestris* L.) and response to fertiliser.

It is usually impossible to identify deficiencies visually in pole-stage crops. There are notable exceptions to this general rule, e.g. where pole-stage crops on infertile sites never received fertiliser or where severe K-deficiency occurs on sites with very high water tables.

Other sampling methods have been tried, including the analysis of the lower crown, bark, litter or soil samples, but none have proved consistently successful. However, biossay of root samples has proved successful for determining P status over a range of crop ages (Dighton and Harrison, 1983). This has been extended to potassium (Jones *et al.*, 1987) and, more recently, to nitrogen. This technique does provide an alternative for determining nutrient status of pole-stage crops and samples would certainly be cheaper to collect. However, the analysis of these three major elements is fairly costly and further work is required to develop the technique fully as a reliable, practical diagnostic tool. If this can be done it would be invaluable for assessing the potential for fertilising spruce crops in the late rotation.

In conifer stands up to 3–4 m high and all Scots pine and broadleaved stands foliar analysis is still the recommended diagnostic aid. Foliar sampling normally takes place at the end of the growing season, in October or November. Deciduous conifers and broadleaves should be sampled in late July or August before needles or leaves begin to change colour (Everard, 1973).

The foliage sampling procedure is detailed in Appendix I. Analysis for nitrogen and phosphorus is done by standard colorimetric methods, following Kjeldahl digestions, and potassium is determined by flame photometry. The results can then be compared with the standard levels in Table 2, although it should be noted that the levels for broadleaves are tentative.

Chapter 3
When to Fertilise

One of the main factors governing growth response to fertiliser application is the stage of development of the crop. For simplicity a crop can be considered as passing through three distinct stages (Miller, H. G., 1981) (see Figure 1).

The first stage occurs between time of planting and the age at which the crop reaches full canopy closure, i.e. when adjacent tree crowns begin to interlock. During this phase the foliage forms a considerable proportion of the total stand biomass and is increasing annually (Madgwick et al., 1977), requiring large amounts of nutrients. These nutrients must come from the pool of available soil nutrients since there is little input from the other possible sources at this stage. Litter fall and nutrient cycling have not fully developed and nutrient input by rainfall is small, the trees being insufficiently developed for effective interception. There is also competition from the ground vegetation for the available nutrients. Therefore any shortfall of nutrients in the soil requires application of fertiliser if the crop is to achieve its full growth potential on the site. It is during this stage that nutrient requirements and growth responses are most predictable and when most forest fertilising takes place in Britain.

The second stage occurs after full canopy closure when recruitment of the new foliage on the tree crowns is offset by shading and death of older foliage in the lowest part of the crown. Therefore, little appreciable increase in foliar mass takes place and the woody parts of the trees form an increasing proportion of the

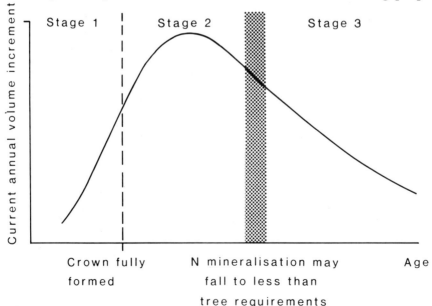

Figure 1. The three distinct nutritional stages in the life of a forest stand, as hypothesised by H. G. Miller (1981).

biomass (Ovington, 1957; Madgwick *et al.*, 1977; Carey, 1980). A proportion of nutrients becomes immobilised in this woody tissue but despite this, most of the crop's demand can be met by nutrients recycled from litter-fall and by the increased crown interception of atmospheric nutrients. Nutrients are also released from the decay of ground vegetation following canopy closure. The demand on the soil nutrient capital is therefore reduced until the crop becomes almost independent of this source of nutrients. Growth responses to fertiliser application are unlikely at this stage, except possibly when the canopy is being restored after thinning operations or defoliation by insects. Fertiliser experiments in pole-stage spruce stands have given very variable responses which could not be related to site factors or foliar nutrient concentrations; when fertiliser was applied in conjunction with thinning the effects were additive and not interactive (McIntosh, 1984a).

The third stage of crop development occurs when nitrogen becomes immobilised in the deepening litter layer (Miller *et al.*, 1976b) to the extent that on sites with a low nitrogen capital the trees begin to suffer nitrogen deficiency. This is usually thought to occur towards the end of long rotations, and is unlikely to be an important problem in British forestry where relatively short rotations are the norm. However, on extremely infertile sites such as the dry heathlands or littoral soils this stage may occur earlier. On such sites good responses to nitrogen application have occurred in pole-stage pine crops (McIntosh, 1984a).

Chapter 4
Where to Apply Fertiliser

When considering where to apply fertiliser it is helpful to divide Britain into three main regions – Scotland and northern England (north of the Humber/Mersey line), Wales, and southern England. The major site types within these regions behave in a similar fashion, although a further breakdown is required within the Scotland and northern England region. The recommendations which follow are derived from experimental results, growth responses to commercial fertiliser applications and surveys of foliar nutrient levels.

Scotland and northern England

Forest areas in this region are mainly extensive plantations of coniferous species. The main species are Sitka spruce and lodgepole pine (*Pinus contorta* Dougl.), with Douglas fir, Scots pine and larches being locally important. It is in this region that virtually all the forest fertilisation programme in Britain is conducted. The soils of this region are very variable and range from fertile soils such as brown earths to extremely infertile soils such as unflushed deep peat (Pyatt, 1982) (see Table 3). Ground water gley soils (group 5), calcareous soils (group 12), ranker and skeletal soils (group 13) and eroded bogs (group 14) have been excluded from Table 3 as they are either not normally planted or form only a minor proportion of upland soils.

The majority of the soil types listed in Table 3 have been placed into four classes, grass-dominated mineral soils (GM), heathland mineral soils (HM), shallow peats (SP) and deep peats (DP). These classes can be regarded as reasonably homogeneous with respect to tree crop requirements for phosphorus and potassium. Each soil type also has one or more category (A, B, C, D) listed which refers to the level of nitrogen deficiency which can be expected in pure Sitka spruce crops (Taylor and Tabbush, 1990).

Plate 10. Phosphorus deficient Sitka spruce crop growing on a grass-dominated brown earth over Tertiary basalt. (R. McIntosh)

The two soil groups, man-made soils (MM) and littoral soils (LS), are very different from the others and they are dealt with in individual sections at the end of this chapter.

Phosphate and potassium fertiliser

Grass-dominated mineral soils

These are principally brown earth, surface-water gley, intergrade and ironpan soils which support a fairly rich ground vegetation consisting mainly of fine grasses, bracken and herbs. They are fertile soils and, although occasional growth responses to phosphate application have been recorded, growth rates are normally well above average. Therefore, other than watching for a possible application of  phosphate in the early thicket stage for species with reasonably high nutrient demands (spruces and firs) fertiliser should not be required (Table 4). Care must be taken when sites have previously been cultivated for agriculture. The resulting disruption to the soil profile can lead to podzols and ironpan soils being classed as brown earths with corresponding overestimates of their inherent fertility.

Weed control can often be beneficial on these sites, for example by removal of early competition from grass (McIntosh, 1983a), and is essential when taller vegetation such as bracken (*Pteridium aquilinum* (L.) Kuhn) is present.

One exception to the above occurs on Tertiary basalts at Fiunary, Mull and Skye (Figure 2). On all soils overlying this lithology, even grass-dominated mineral soils, there is very little available phosphate (Plate 10). This may be linked to high exchangeable aluminium levels which lead to phosphate becoming fixed within the soil. However, it is evident from experiments that regular applications of phosphate can overcome this problem. On these lithologies, grass-dominated mineral soils should have the same inputs of phosphate as heathland mineral soils. The interval between applications should be 6 years.

1. Lewisian / Torridonian
2. Tertiary Basalt
3. Moine
4. Upper / Middle Old Red Sandstone
5. Dalradian
6. Lower Old Red Sandstone
7. Carboniferous *
8. Silurian / Ordovician
9. Jurassic **
10. Cambrian
11. Hastings Beds

* Includes Millstone Grit and Pennant Sandstone

** Includes Estuarine gleys

Figure 2. Simplified map of main lithologies mentioned in the text (for a more detailed breakdown, refer to British Geological Survey maps).

Table 3. Forestry Commission soil classification (after Pyatt, 1982) with additional classes (Taylor, 1986) and categories (Taylor and Tabbush, 1990)

Soil group	Soil type	Code	Class*	Category
1. Brown earths	Typical brown earth	1	GM	A
	Basic brown earth	1d	GM	A
	Upland brown earth	1u	GM	A B
	Podzolic brown earth	1z	HM	A B
	Ericaceous brown earth	1e	HM	A B C
3. Podzols	Typical podzol	3	HM	B C D
	Peaty podzol	3p	SP	B C
4. Ironpan soils	Intergrade ironpan soil	4b	GM(HM)	A B C
	Ironpan soil	4	GM/HM	A B C D
	Podzolic ironpan soil	4z	HM	B C D
	Peat ironpan soil	4p	SP	A B C
6. Peaty gley soils	Peaty gley	6	SP	A B C D
	Peaty podzolic gley	6z	SP	B C
7. Surface-water gley soils	Surface-water gley	7	GM	A B C
	Brown gley	7b	GM	A
	Podzolic gley	7z	GM/HM	A B C
8. *Juncus* bogs (basin bogs)	Phragmites bog	8a	DP	A
	Juncus articulatus or *acutiflorus* bog	8b	DP	A
	Juncus effusus bog	8c	DP	A
	Carex bog	8d	DP	A
9. *Molinia* bogs (flushed blanket bogs)	*Molinia, Myrica, Salix* bog	9a	DP	A
	Tussocky *Molinia* bog; *Molinia, Calluna* bog	9b	DP	A B
	Tussocky *Molinia, Eriophorum vaginatum* bog	9c	DP	B C
	Non-tussocky *Molinia, Eriophorum vaginatum, Trichophorum* bog	9d	DP	B C
	Trichophorum, Calluna, Eriophorum, Molinia bog (weakly flushed)	9e	DP	B C D
10. *Sphagnum* bogs (flat or raised bogs)	Lowland *Sphagnum* bog	10a	DP	D
	Upland *Sphagnum* bog	10b	DP	D
11. *Calluna, Eriophorum, Trichophorum* bogs (unflushed blanket bogs)	*Calluna* blanket bog	11a	DP	C D
	Calluna, Eriophorum vaginatum blanket bog	11d	DP	C D
	Trichophorum, Calluna blanket bog	11c	DP	D
	Eriophorum blanket bog	11d	DP	D
2. Man-made soils	Mining spoil, stony or coarse textured	2s		
	Mining spoil, shaly or fine textured	2m		
15. Littoral soils (coastal sand/gravel)	Shingle	15s		
	Dunes	15d		
	Excessively drained sand	15e		
	Sand with moderately deep water table	15i		
	Sand with shallow water table	15g		
	Sand with very shallow water table	15w		

Classes: GM = grass-dominated mineral soils, HM = heathland mineral soils, SP = shallow peats, DP = deep peats.

Table 4. Recommended regimes of phosphorus and potassium

Soil category	Species*	*Crop age (years)		
		0†	6-8	12-16
Grass-dominated mineral soils	SS, NS, DF	—	(P)	—
	SP, LP, L, BL	—	—	—
Heathland mineral soils	SS, NS, DF	P	P	(P)
	SP‡, LP, L, B/L	(P)	(P)	—
Shallow peat (<45 cm)	SS, NS	P[K]	P[K]	—
	SP, LP, L, B/L	(P[K])	(P[K])	—
Deep peat (>45 cm)				
Flushed *Juncus* bogs	SS, NS, BL	—	—	—
Flushed *Molinia* bog	SS	PK	PK	—
	LP	(PK)	(PK)	—
Unflushed bogs	SS	PK	PK	(PK)
	LP	PK	PK	—

* SS = Sitka spruce, NS = Norway spruce, DF = Douglas fir, SP = Scots pine, LP = lodgepole pine, L = Japanese, European and hybrid larch, BL = broadleaves.
† This refers to the date the crop is planted. There is a 6 month tolerance either side of this date for the application of initial fertiliser to be fully effective.
‡ Topdressing of Scots pine in the pole-stage can also be beneficial on heathland.
( ) indicates possible benefit.
[ ] indicates that potassium will be required when the peat depth exceeds 30 cm.

Heathland mineral soils

This category is dominated by podzols, ironpan soils and podzolic gleys which occur mostly in the drier, eastern half of the region. The main component of the ground vegetation is heather (*Calluna vulgaris* (L.) Hull) and availability of phosphate is lower than in grass-dominated mineral soils. Species such as Sitka spruce, Norway spruce and Douglas fir require phosphate at planting and at between 6 and 8 years in the early thicket stage (Table 4). On the most phosphate-deficient lithologies (Taylor and Worrell, 1991a), such as Moine schists, Dalradian quartzite, Tertiary basalt lava and Jurassic (Estuarine) shales (Figure 2), the response period is likely to be only 6 years instead of the normal 8 and a third application of phosphate may be necessary. Potassium deficiency is not a problem on heathland mineral sites.

Less demanding species such as Scots pine, larch or birch will grow satisfactorily on most heathland sites without fertiliser. However, an application of phosphate at planting and again in the early thicket stage is beneficial on the poorer lithologies listed above. Lodgepole pine is also responsive to application of phosphate at planting on any heathland site and to a top dressing in the early thicket stage on the poorer lithologies (Table 4). A similar prescription is appropriate for broadleaves.

Application of phosphate to pole-stage Scots pine crops on heathland sites has also proved to be beneficial (McIntosh, 1984a). This occurred where foliar phosphorus levels fell below 0.2%. It is worth noting that this level is much higher than that identified as marginal for young crops by Binns *et al.* (1980).

Shallow peats (< 45 cm)

These are mainly peaty gleys and peaty ironpan soils where the vegetation includes *Calluna*, either dominant or in mixture with *Molinia* and *Trichophorum*. The fertility of these soils depends both on the lithology and the depth of peat (range of 5 to 45 cm). Phosphorus deficiency often limits growth and potassium generally becomes deficient when peat depth exceeds

30 cm (Taylor, 1986). Demanding species (Sitka and Norway spruce) require application of phosphate at planting and again in the early thicket stage (Table 4). On the most phosphate-deficient lithologies, such as Moine schists, Tertiary basalt lava and Jurassic shales (Figure 2), the response period is likely to be only 6 years instead of the normal 8 and a third application of phosphate may be necessary. Whenever the peat depth exceeds 30 cm potassium should be applied in addition to phosphate. This is particularly important on lithologies low in available potassium, such as Old Red sandstone (Upper and Middle), Carboniferous sediments, Basic Igneous, Silurian and Ordovician shale (Figure 2). On Moine schists (Figure 2) potassium deficiency is often not encountered at time of planting and application of potassium can be delayed until the first top dressing of phosphate is applied after 6 years.

Less-demanding species (pines and larches) respond to application of phosphate and potassium, particularly on the poorer lithologies. However, even on the poorest lithologies no more than two applications of fertiliser are required (Table 4).

Deep peats (> 45 cm)

This category comprises all soils with an organic horizon greater than 45 cm and contains a considerable range of peat types. The Forestry Commission soil classification (Pyatt, 1982) recognises 18 types but these can be grouped into three main fertility classes – strongly flushed *Juncus* bogs, flushed *Molinia* bogs and unflushed bogs.

Juncus bogs

Juncus bogs (group 8 in Table 3) are strongly flushed and therefore quite fertile due to the continual movement of water and nutrients through the soil. They are capable of supporting high yields with no fertiliser input.

Molinia bogs

The *Molinia* bogs (group 9 in Table 3) have a lesser degree of flushing and are less fertile. Crops require phosphate and potassium applications (Table 4), particularly when Sitka spruce has been planted (requirements for lodgepole pine are not as great). These requirements increase as species such as *Eriophorum vaginatum* L., *Trichophorum caespitosum* (L.) Hartman and *Calluna* form an increasing percentage of the ground vegetation or where the soils occur on Moine schist, Tertiary basalt and Jurassic shale (Figure 2).

Unflushed bogs

The unflushed bogs (groups 10 and 11 in Table 3) are very infertile soils where two applications of phosphate and potassium are normally required on Sitka spruce and lodgepole pine crops in the establishment phase (Table 4). A third application is required for Sitka spruce on Moine schist and Tertiary basalt (Figure 2) or where potassium deficiency could be a serious problem (Upper/Middle Old Red sandstone, Basic igneous, Carboniferous sediments, Silurian and Ordovician shales – see Figure 2). Again, on Moine schists potassium deficiency may not be encountered at planting. However on these unflushed peats it often occurs within 4 years and it is more sensible to apply potassium at planting to prevent growth check rather than to wait until the first top dressing.

There is obviously a considerable range of fertility in the deep peat class but, within the main groupings indicated, there are reasonably similar fertiliser requirements.

Nitrogen fertiliser

On peatland and heathland soils, nitrogen-deficiency can severely restrict the growth of Sitka spruce, the main commercial species. Douglas fir, Norway spruce and silver firs (*Abies* spp.), although not commonly planted on these sites, also encounter the same problem. Until the 1970s it was thought that nitrogen deficiency was caused by competition from heather (Fraser, 1933; Weatherell, 1954). Commonly called 'heather check', this was later identified as an allelopathic effect (Malcolm, 1975), i.e. caused by the production of antagonistic chemicals which inhibit root growth. However, increased planting of Sitka spruce on very nutrient-poor soils revealed that, even after removal of heather by herbicide treatment, growth was still limited by low availability of nitrogen (Mackenzie, 1974). This can be caused by limited soil nitrogen capital and/or slow rate of nitrogen mineralisation (McIntosh, 1983b). Application of nitrogen fertiliser can overcome this deficiency although several applications may be required to achieve full canopy closure (McIntosh, 1983b). Once this stage is reached demand for nutrients is reduced due to shading of competing vegetation, improved nutrient cycling and capture of atmospheric nutrients (Miller, H. G., 1981) and further inputs of nitrogen should not be required.

Pines and larches may be marginally nitrogen-deficient on occasion but their growth is not significantly checked. However, pole-stage fertilising of Scots pine crops on heathland sites has proved to be beneficial (McIntosh, 1984a). Below a threshold foliar nitrogen level of 1.8%, the response to nitrogen application increased as the foliar concentration dropped. This threshold level again is much higher than that identified for young crops by Binns et al. (1980).

Categorisation of nitrogen-deficient sites
The major difficulty facing forest managers in determining the treatment of a nitrogen deficient stand is deciding whether heather control, application of nitrogen fertiliser or a combination of both will yield the most cost-effective response on any given site. To solve this problem a system has been developed for dividing nitrogen sites carrying susceptible crops into four categories (Taylor and Tabbush, 1990):

Category A
Here there is sufficient nitrogen available for acceptable tree growth, despite the presence of heather. The inhibitory effect of heather seems to be reduced when soils are rich in available nitrogen, and susceptible species are unlikely to suffer any real check to growth, although there may be a slight yellowing of foliage in the 2 or 3 years prior to canopy closure. Normally these are sites where the heather is mixed with fine grasses, such as *Agrostis*, *Festuca* and *Anthoxanthum* (spp.), in the transition zone between grassland and heathland; or weakly flushed peats dominated by *Myrica* and vigorous *Molinia*; or sites heavily colonised by broom (*Cytisus scoparius* (L.) Link) or gorse (*Ulex europaeus* L.). No herbicide or fertiliser is required.

Category B
The sites in this category are those where heather is the principal cause of nitrogen deficiency, and successful heather control with 2,4-D or glyphosate (Williamson and Lane, 1989) results in adequate availability of nitrogen for susceptible species. These are usually heathlands on more fertile lithologies (e.g. basic igneous, phyllites, pelitic schists) or western *Molinia/Eriophorum* peatland where the heather is sub-dominant.

Category C
Heather is the dominant type of vegetation on these sites, but is not the sole cause of nitrogen deficiency. The low mineralisation rate is also a major factor and although heather control results in a cost-effective growth response, it will not bring permanent relief from nitrogen deficiency and subsequent inputs of nitrogen fertiliser will be required to achieve full canopy closure. This category can include peats where *Molinia* and *Trichophorum* are co-dominant with *Calluna* and certain heathland soils with low organic matter content.

Category D
The principal cause of nitrogen deficiency on these sites is the low mineralisation rate. Heather control does not give a cost-effective growth response. In fact, on many of these sites heather is either not present or very sparse. Nitrogen fertiliser has to be applied every 3 years to Sitka spruce (McIntosh, 1983b) from onset of deficiency until full canopy closure is achieved, meaning that three to five inputs are required. This category contains lowland and upland raised bogs but it includes podzolic soils with low organic matter on quartzitic drifts.

The various categories for each soil type are listed in Table 3. When identifying the exact category for a particular site the following pro-

Table 5. Ranking of the main lithologies according to the likely availability of nitrogen in overlying soils

Group I Low nitrogen availability	Geological map* reference numbers
Torridonian sandstone	61
Moine quartz-feldspar-granulite, quartzite and granitic gneiss	8, 9, 10, 12
Cambrian quartzite	62
alradian quartzites	17
Lewisian gneiss	1
Quartzose granites and granulites	34 (part only)
Middle/Upper Old Red Sandstone (Scotland)	77, 78
Upper Jurassic sandstones and grits	97, 98, 99
Carboniferous grits and sandstones	81 (part only)
Group II Moderate nitrogen availability	
Moine mica-schists and semi-pelitic schists	11
Dalradian quartzose and mica schists, slates and phyllites	18, 19, 20, 21, 23
Granites (high feldspar, low quartz content)	34 (part only)
Tertiary basalts	57
Old Red Sandstone basalts, andesite and tuff	44, 46, 47, 48, 50
Silurian/Ordovician greywackes, mudstones (Scotland)	70, 71, 72, 73, 74
Lower and Middle Jurassic sediments	91, 94, 95
Group III High nitrogen availability	
Gabbro, dolerite, epidiorite and hornblende schist	14, 15, 26, 27, 32, 33, 35
Lower Old Red Sandstone	75
New Red Sandstone	85, 89, 90
Carboniferous shales and basalts	53, 54, 80**, 81 (part only), 82, 83, 84
Silurian/Ordovician/Devonian shales (Wales and south-west England)	68, 69, 70, 71, 72, 73, 74, 75, 76, 77, 78
Limestones	24, 67, 80***, 86
Cambrian/Precambrian	60, 64, 65, 66

*   Reference: Institute of Geological Sciences Geological Map of the United Kingdom (3rd edition *solid,* 1979), published by the Ordnance Survey.
**   refers to Scotland only.
***   refers to England and Wales only.

Notes: 1. Geological Map index no. 34 has been subdivided into: (a) quartzose granites and granulites (Group I), and (b) granites with a high feldspar and low quartz content (Group II).
2. Geological Map index no. 81 has been subdivided into: (a) grits and sandstones (Group I), and (b) shales (Group III).
3. Where soils occur over drift material, then their characteristics (in terms of nitrogen availability) will be similar to that of the solid parent material from which the drift was derived.

cedure should be used (Taylor and Tabbush, 1990). First find the appropriate soil type in Table 3 and commence from the left hand side of the categories listed against that soil type. If more than one category is listed, identify the appropriate lithology group in Table 5. If this lies within group I, then move two categories to the right; within group II move one category to the right; within group III stay in the same category. Finally, if the soil type is mineral or organo-mineral (soil groups 1, 3, 4, 6, or 7 in Table 3) and the site is dominated by *Calluna* (more than 50% ground cover – equivalent to the 'ericaceous phase' mapped in Forestry Commission soil surveys) move one category to the right. If not, then stay in the same category.

Once the category has been determined for a particular site the manager has a clear idea of the best action for treating a nitrogen deficient crop. Problems identified on category A sites are temporary and should not require treatment. The most cost-effective solution on category B sites is to control the heather, whereas on category D sites a programme of nitrogen application is advised. The choice is most difficult on category C sites where there will normally be a good response to heather control but subsequent applications of nitrogen are required. In many cases it is preferable to control the heather first but, if weather or logistical problems prevent this, then there is unlikely to be a large difference in cost should the manager decide to invest only in a programme of nitrogen application.

Correct choice of species and nursing mixtures

On category C and D sites, slower growing and less demanding species, such as Scots pine, lodgepole pine and larch can be used as a low input/low output option. However, they can also be planted as a nurse in mixture with Sitka spruce to prevent the latter suffering a prolonged period of nitrogen deficiency (Taylor, 1985). It is now recognised that this nursing effect is not only due to suppression of the heather, but that there is also a considerable increase in nitrogen availability (O'Carroll, 1978; Carlyle and Malcolm, 1986). The mechanisms behind this increase are still not fully understood despite recent intensive studies in Britain and Ireland (Carey *et al.*, 1988). There are indications that the spruce receives the nitrogen via the nutrient cycle of the mixed crop or via mycorrhizal associations (symbiotic root fungi) in which the nurse species is involved (I. Alexander, personal communication), and that nitrogen demand is reduced by diluting the amount of Sitka spruce canopy by establishing a mixed stand (H. G. Miller, personal communication). Whatever the mechanism, the evidence from experiments and field experience is sufficiently convincing to support the use of this technique in general practice (Carey *et al.*, 1988; Taylor and Tabbush, 1990).

Plate 11. *The better growth of the Sitka spruce growing in mixture with Scots pine on the left compared with the pure Sitka spruce on the right (both 20 years old) is due to the nursing benefit of the pine on this infertile heathland site.* (C. P. Quine)

Nurse species can be used on both the heaths and the peatlands although species choice and mixture pattern may be different (Taylor, 1985). On the heathlands Scots pine has proved a very satisfactory nurse (Plate 11), but planting stock from a suitable seed origin must be used to ensure good survival and growth (Lines, 1987). Japanese larch is also a very effective nurse but can be an aggressive competitor in the thicket and pole-stage. On the peatlands the best choice is lodgepole pine which not only nurses the Sitka spruce but is a very effective pioneer on these sites. Alaskan or North Coastal provenances are the most suitable choice (Lines, 1987) as they establish well but do not grow too rapidly in the early years (Plate 12).

Current recommendations are to have a mixture ratio of 1:1, planted in a fairly intimate mixture at normal spacing (Taylor, 1985). In most cases the pattern should be alternate pairs or triplets in every row, planted in a staggered fashion. This ensures proximity of Sitka spruce to the nurse but gives some insurance against suppression, should the nurse prove more vigorous than expected. There is good historical evidence to support row mixtures particularly for Scots pine/Sitka spruce mixtures on dry, heathland sites. On sites of low windthrow hazard it is then possible to remove the nurse by line thinning before it is suppressed. However, there may be some danger in removing the nurse before it has completed its task on the most infertile sites.

There are two situations where the use of nursing mixtures will not be successful. Where there is a combination of a heathland site over quartzite and low rainfall likely to cause moisture stress in Sitka spruce, or where there is likely to be a high incidence of frost damage to Sitka spruce, pure pine should be planted.

It is also possible to use nursing mixtures as a rehabilitation treatment by interplanting severely checked Sitka spruce crops with a nurse species. This has been successfully used in the past but there are several drawbacks to this technique. It could be difficult and costly to establish nurse species due to vulnerability to weed competition and browse. There is also a delay of approximately 10 years before the benefit arises. Therefore, it is unlikely that this practice will be widely used, although it should be considered where pure Sitka spruce has been very recently planted on sites where severe nitrogen deficiency can be expected and beating-up programmes are planned.

Nitrogen-fixing plants

Nitrogen-fixing plants are used successfully in other countries for nursing (Davey and Wollum, 1984), and experimentation in Britain with leguminous plants, particularly broom, has demonstrated their nursing benefits (Nimmo and Weatherell, 1961). Despite this success, the technique has never been adopted commercially

Plate 12. *The good vigour and high nitrogen status of this Sitka spruce (18 years old) growing on an unflushed deep peat is because of the nursing benefit derived from being in mixture with Alaskan lodgepole pine.* (C. M. A. Taylor)

in practice due to the difficulty of successfully establishing the broom and protecting it from browsing. Therefore it is unlikely to be a dependable alternative to using conifer nurses but should be utilised where it occurs naturally.

Alders (*Alnus* spp.) are another possible nurse and certainly fix considerable quantities of nitrogen (Gessel and Turner, 1974), which would be more than enough to prevent nitrogen deficiency in Sitka spruce. However, there are difficulties in establishing alders on poor upland sites (particularly peatlands) which militate against their widespread use at present (Lines and Brown, 1982).

Sewage sludge

The use of organic wastes containing high levels of nitrogen has potential on nitrogen deficient sites. Sewage sludge is a readily available source of nitrogen which is currently used on agricultural land in Britain and has been successfully used elsewhere in forestry (Bastion, 1986; Nichols, 1989). Similarly, effluent from distilleries and breweries has been successfully utilised in the past for reclamation of heathland to agriculturally-valuable grassland. Current research indicates that application of sewage sludge has markedly improved the growth of Sitka spruce on a heathland afforestation site in the early years (Bayes *et al.*, 1989) and eliminated the heather, although the duration of response has still to be determined. Other experiments are testing its value on pre-thicket and pole-stage pine crops (Plate 13) and on heathland restocking sites.

Where supplies of sludge are available within 10-20 miles and there is good access to

Plate 13. *Application of liquid sludge by modified irrigation equipment to a pole-stage pine crop.* (P. M. Tabbush)

and within the forest, this option has great potential, offering a long-term improvement in the nitrogen availability on well-drained heathland sites (Taylor and Moffat, 1989). It is unlikely to be suitable for wet, peaty sites where there is a risk of run-off. The use of sewage sludge in forests is still in the development phase in Britain and pilot scale trials are being established under a draft Code of Practice (Bayes and Taylor, 1989).

Application of lime
Applying lime raises the pH of the soil and increases microbial activity, thereby increasing the rate of nitrogen mineralisation in the long term. Experimental results in Britain, however, have not been very encouraging (McIntosh, 1983b), although work in Northern Ireland has indicated long-term benefits in nitrogen availability and tree growth (Dickson, 1977). Lime is costly and difficult to apply at the heavy rates required (5–10 tonnes ground limestone per hectare), therefore this treatment cannot be presently recommended.

Wales

Wales has representatives of all the major soil types found in Scotland and northern England, but the restricted lithological range reduces variability in soil nutritional status. Deep peats and heathland soils are not extensive, and in general the soils are more fertile than in the northern region. The area is also climatically more favourable and the higher soil temperatures may allow a more rapid turnover of nutrients. Currently, the annual fertiliser programme in Wales is less than 3% of the total programme in Britain.

Growth responses may be achieved on the heathland mineral soils from herbicide control of the heather (for species susceptible to check) and a phosphate application at around year 8 (Everard, 1974). Crops growing on *Molinia*-dominated soils overlying Silurian and Ordovician shales, Cambrian and Millstone grit (Figure 2) benefit from a phosphate application at planting, although this is not essential, and at between 8 and 15 years (Everard, 1974).

Potassium should also be applied in conjunction with phosphate on these sites where peat depth exceeds 30 cm. On the deeper hill peats nitrogen application may also be beneficial to pure Sitka spruce in the thicket stage. On the unflushed bogs an application of phosphate and potassium is required at planting; these nutrients plus nitrogen should be applied to pure Sitka spruce crops in the thicket stage at year 8–10. There are recent indications that Sitka spruce on peaty gleys and deep peats over Pennant sandstone in the South Wales coalfield (see Figure 2) is responding to application of nitrogen, phosphate and potassium in the thicket stage. It is hoped this might help prevent the growth decline in spruce that has been observed in this area (Coutts *et al.*, 1985).

Southern England

This region contains a higher proportion of fertile soils than the other two regions, hence lack of nutrients does not generally limit tree growth and very little fertiliser is applied. Broadleaves, such as oak (*Quercus* spp.), beech (*Fagus sylvatica* L.), poplar (*Populus* spp.), sweet chestnut (*Castanea sativa* Mill.), elm (*Ulmus* spp.) and ash (*Fraxinus excelsior* L.), form a significant proportion of the forest area. The main conifer species are Scots pine and Corsican pine (*Pinus nigra* var. *maritima* (Ait.) Melville), although Sitka spruce, Douglas fir, larch and Norway spruce are locally important.

Phosphate should be applied in the establishment phase of crops planted on the poor heathland or former heathland soils of south-west England (Cornwall, Devon, Somerset, Dorset and western Hampshire) and on the Hastings Beds in south-east England (Everard, 1974). Heather is also present on the drier heathlands but, as pines are the main species planted on these soils, it is not a major problem. The only other cases where potential growth benefits to fertiliser application have been suggested are with nitrogen on pole-stage ash crops (Evans, 1986a) and phosphate on sweet chestnut coppice (Evans, 1986b). However, neither of these responses has been fully proven and they are not recommended practice at present.

Restocking

The preceding sections in this chapter have outlined where fertiliser application is beneficial, but refer only to first rotation crops. Therefore some reference should be made to the probable nutrient requirements of second rotation crops as an increasing proportion of the forests mature and are felled. The current rate of clear felling in Britain's state and private forests is approximately 9000 hectares a year, and will almost double in the next 20 years (Taylor, 1990b).

A recent series of Sitka spruce nutrition experiments established on a range of restocking sites indicates that replanted crops have a higher nutrient status than first rotation crops on the same sites (Taylor, 1990b) (Plate 14). The main reason for this is the decomposition of the brash left after harvesting (needles, branches and tops) and the litter layer (Titus and Malcolm, 1987). However, there are indications that nutrient leaching from the site (Malcolm and Titus, 1983) could result in deficiency problems in the late thicket stage when the crops are making maximum demands on the site.

Concern has been expressed that more intensive harvesting systems might create nutrient deficiency problems. Carey (1980) found that removal of Sitka spruce brash could increase biomass yield by 20% but would also increase removal of nitrogen by 240%, phosphate by 330% and potassium by 180%. This impact could be halved by delaying removal until needles had fallen. However, these large figures only represent a small proportion of the total nutrient store on most sites and this practice would only give serious concern on the poorer sites, particularly where rotations may be shortened by other site factors.

Although these investigations are at an early stage it is highly likely that those crop types and soils where fertiliser was not needed in the first rotation will not require it in the second rotation. It is also fairly clear that restocked crops do not require fertiliser at planting, and inputs over the whole rotation may be substantially reduced. One exception occurs on non-peaty heathland sites (particularly those listed as category D sites) where pure Sitka spruce crops are suffering from severe nitrogen-deficiency, following previous crops of pure pine or larch (Taylor and Tabbush, 1990). These crops require several inputs of nitrogen to achieve full canopy closure. It is recommended that such sites should be replanted with pure pine or larch, or a mixture of Sitka spruce and Scots pine.

Another possible exception has been identified following foliar analysis of Sitka spruce crops on Estuarine peaty gleys in the North York Moors. Deficient phosphorus levels occur on these sites within 5 years of planting and it is likely that an application of phosphate will be required. However, this is not entirely surprising as this site type is known to be very low in available phosphorus (Taylor and Worrell, 1991a).

Plate 14. *Vigorous Sitka spruce (12 years old) on an unfertilised restock site on unflushed deep peat with no indication of nutrient deficiency.* (C. M. A. Taylor)

Unusual sites

Man-made soils

There is an increasing interest in planting trees on reclaimed land, either following extraction of minerals (e.g. coal, oil shale, gravel) or on derelict industrial sites. The major nutritional problem associated with such sites is nitrogen-deficiency, usually due to a lack of organic matter. This can be dealt with by choosing appropriate tree species (e.g. alder, birch) to try and reduce the problem and planting nitrogen-fixing species (Binns and Fourt, 1981), such as broom, lupins (*Lupinus* spp.) and clovers (*Melilotus* and *Trifolium* spp.). Another option is organic amendment with sewage sludge (Bayes *et al.*, 1989) or peat (McNeill and Moffat, 1989), which may be the best approach in exposed upland sites where species choice is limited. Many of these sites are near sludge sources or will have had peat stripped off prior to excavation.

Applications of inorganic nitrogen would also correct nitrogen deficiency where unsuitable species have been planted, but these are likely to be prohibitively expensive because of the number of applications required. The options described above offer a more realistic solution.

Inorganic phosphate fertiliser is often needed on such sites (Jobling and Stevens, 1980). When soil pH is near neutral, soluble phosphate fertilisers (e.g. superphosphate) are a better source than rock phosphate. Where sewage is applied to improve nitrogen availability this will also supply sufficient phosphate for crop requirements.

Littoral soils

Large forest blocks have been established on windblown sand and dune systems in several coastal locations, such as Culbin, Roseisle, Tentsmuir, Newborough and Pembrey. The major nutritional problem with these sites is lack of nitrogen associated with low levels of organic matter. These areas were mainly planted with Corsican and Scots pine which have relatively low demands for nitrogen. However, even these species become nitrogen deficient in the pole-stage on some of these soils (Wright and Will, 1958) and Corsican pine in particular shows large responses to application of nitrogen fertiliser (Miller *et al.*, 1976b).

Applications of inorganic nitrogen fertiliser can be cost effective by boosting yield in the latter part of the rotation. The use of leguminous nurse species also offers a practical alternative on these low-lying sites. Undersown lupins have been very effective in an experiment in Culbin, increasing the basal area by 32% over a 20-year period (Plate 15). If this method is chosen it is imperative that the treated area is protected from grazing. Another alternative is to use organic fertilisers, such as sewage sludge, which not only provide nitrogen but also improve soil structure and water retention.

High elevations

A series of experiments has been planted at high elevations to compare the effects of a range of fertiliser inputs on the establishment of Sitka spruce and lodgepole pine. This has demonstrated that the application of fertiliser, even at luxurious rates, is ineffective when climate becomes the major limiting factor to growth. This should be borne in mind when determining upper planting margins or when contemplating topdressing of crops which are suffering badly from exposure.

Plate 15. *Pole-stage Scots pine growing on sand dunes showing the benefits of underplanting with lupins (area beyond the fence).* (R. McIntosh)

Chapter 5
Types and Rates of Fertiliser

A fertiliser for use in forestry must be cheap, sufficiently concentrated for low application costs and be in a suitable form for aerial and ground application. It should also be relatively long-lasting and unlikely to contaminate the environment beyond the forest.

Phosphorus

The various types of phosphorus fertiliser (Binns, 1975) can be divided into two main groups, water-soluble and water-insoluble.

Soluble vs. insoluble

Two water-soluble fertilisers have been tested – single superphosphate and triple superphosphate (Zehetmayr, 1954; Edwards, 1958). Single superphosphate performs reasonably well but has a high cost per unit of phosphorus and a relatively low concentration (18–21% total P_2O_5), making application costs high. Triple superphosphate has a much higher concentration (44–47% total P_2O_5) which keeps application costs down but is expensive per unit of phosphorus. It is now only used in circumstances where it is crucial to reduce the weight of fertiliser applied, e.g. hand applications in remote areas.

Two water-insoluble phosphorus fertilisers have also been evaluated – basic slag (by-product of steel making) and rock phosphate, both of which are much cheaper than the water-soluble fertilisers (Binns, 1975). For many years basic slag was used almost exclusively but the modernisation or closure of many steel-making plants meant supplies became scarce. Furthermore the excessive fineness of the material and low concentration (7–22% total P_2O_5) caused application difficulties and it is no longer used.

Rock phosphate has now become the main source and has proved to be as effective as the superphosphates, being soluble in the acid conditions prevalent in most forest soils (McIntosh, 1984b). However, the phosphorus content and acid solubility of the common rock phosphate sources varies, with total P_2O_5 content ranging from 27% to 39% and citric acid solubility from 3% to 13.5% (Binns, 1975). Good growth responses have been obtained with the sources which have a minimum P_2O_5 content of 27% with at least 9% of the material being soluble in citric acid (Binns, 1975; McIntosh, 1984b).

The first rock phosphate tested was ground mineral phosphate which largely replaced basic slag in the 1930s and 1940s. However, this finely ground rock phosphate was found unsuitable for aerial applications, clogging up the spreaders and proving too vulnerable to wind-drift. Unground rock phosphate then became the preferred fertiliser, particularly the North African sources with their relatively high acid solubilities. This slightly coarser material proved more suitable for aerial application without reducing growth responses (Mackenzie, 1972) and it is less prone to leaching from very acid peat soils with good drainage (Malcolm *et al.*, 1977; Ballard, 1984). However, there is still a high proportion of fine material (up to 80% will pass through a 0.25 mm sieve). Granulated rock phosphate has been shown to improve the distribution patterns of aerial application (Farmer *et al.*, 1985) and, provided a suitable source of rock phosphate is used, is an equally good source of phosphorus. The main drawback is that granular fertiliser costs 30–40% more per tonne than unground rock phosphate, although because subsequent handling and application costs are fixed this is reduced to 10–20% per treated hectare.

Rate and season of application

The current application rate for rock phosphate is 450 kg per treated ha (to supply 60 kg phosphorus per treated ha) which is a compromise between cost of application and effectiveness (McIntosh, 1984b). To offset the cost of granular fertiliser it would be possible to reduce the rate to 400 kg per treated ha on the more fertile lithologies. This is in anticipation that the reduction in potential growth rate is compensated by more even aerial application.

The period of growth response is 6–10 years depending on soil type, tree species and lithology (see Chapter 4), although there is evidence that the period may be longer on grass-dominated mineral soils (Taylor and Worrell, 1991b). The season of application has no real impact on the effectiveness of the fertiliser provided that it is not applied when the ground is frozen or snow-covered (McIntosh, 1984b).

Potassium

Types

There are two main sources – muriate of potash (potassium chloride) and potassium sulphate (Binns, 1975). Muriate of potash is almost exclusively used because it is cheaper, more widely available and is more concentrated (60% K$_2$O against 50%). Although it is a readily soluble fertiliser there is normally a 6–8 year growth response (McIntosh, 1981), probably due to the ability of trees to store excess potassium for future translocation when required (Miller, 1984). Another material, adularia shale, has recently been tested. This is an orthoclase mineral (part of the feldspar group) which contains 10–12% K$_2$O and it was hoped it would act as a slow release fertiliser. However, it has not proved to be an effective source of potassium for forest use (Dutch *et al*., 1990).

Rate and season of application

The currently recommended application rate is 200 kg muriate of potash per treated ha (to supply 100 kg potassium per treated ha). Season of application has no great effect on the growth response (Dutch *et al*., 1990). However, the normal precaution of avoiding application to snow-covered or frozen ground must be followed.

Potassium fertiliser is nearly always applied in conjunction with phosphorus fertiliser. Previously, this was a straightforward mix of unground rock phosphate and muriate of potash (i.e. 650 kg per treated ha) where it is important that the components are uniformly mixed and have a similar range of particle sizes to prevent separation or settlement during transit or distribution (Binns, 1975). Increasingly, there is greater use of compound fertilisers where these two components are granulated (see p. 25).

Nitrogen

The most commonly used nitrogen fertilisers in forestry are **ammonium nitrate** and **urea** (Binns, 1975), both having high nitrogen content (34% and 46% respectively) and being water soluble. They give a good compromise between the cost of the fertiliser and low application rate and handling costs.

Urea vs. ammonium nitrate

There is conflicting evidence over which material is the more effective source of nitrogen. Some experimental results from Scandinavia (Moller, 1974), Canada (Dangerfield and Brix, 1981; Weetman and Algar, 1974) and the northwest of the United States (Harrington and Miller, 1979) suggest that ammonium nitrate is better. Other work in the United States (Ballard, 1981; Fisher and Pritchett, 1982; Radwan *et al*., 1984), Ireland (Griffin *et al*., 1984) and Great Britain (Taylor, 1987) has indicated no difference between the two in growth response. Poorer responses with urea have been associated with high volatilisation losses of ammonia (Otchere-Boateng, 1981). However in the cool, humid climate of Britain volatilisation losses are likely to be much lower than in a more continental climate (e.g. Scandinavia).

Urea must first be converted to ammonium and then to nitrate, during which process ammonium becomes immobilised in the organic layers (Overrein, 1970). This process is also slowed down in wet, cold soils (Malcolm, 1972). However, these factors can be regarded as advantageous in upland soils prone to heavy leaching, where application of ammonium nitrate could result in nitrate contamination of watercourses (Tamm et al., 1974).

Urea is more frequently used in British forestry since it is generally cheaper than ammonium nitrate. This is not always the case and relative prices should be checked before ordering fertiliser.

The only case where ammonium nitrate is recommended in preference to urea is for topdressing of pole-stage Scots pine (McIntosh, 1984a).

Rate and season of application

The current recommended rate is 150 kg elemental nitrogen per treated ha which is rounded up to 350 kg urea (or 440 kg ammonium nitrate) per treated ha (McIntosh, 1983b). This rate is a compromise between fertiliser cost and growth response as responses continue on a 'law of diminishing returns' above this rate. The normal response period to any application of nitrogen is only three growing seasons (McIntosh, 1983b) so that nitrogen deficient crops on category C and D sites must be treated on a 3-year cycle until full canopy closure occurs and nutrient cycling commences.

In the past, the season of application was thought to be critical and treatment was confined to late spring and early summer. Recent evidence, however, indicates that timing is not critical in Britain, possibly as a result of the mild, maritime climate with no lengthy periods of extreme weather (Taylor, 1987). Therefore, provided periods subject to snow or frost are avoided (due to risk of run-off), there is no disadvantage in applying nitrogen towards the end of, or even outwith, the growing season. However, increment in that year will be lost (recouped at the end of the response period) if application is carried out after June. Late applications (August onwards) should be avoided in upland valleys subject to early frosts.

Other nitrogen sources

Other inorganic fertilisers which have been tested are slow-release and liquid nitrogen fertilisers (Fisher and Pritchett, 1982; Mead et al., 1975; Miller, R. E., 1981). These are potentially beneficial through improving the efficiency of nitrogen uptake by the trees, which is often less than 30% in conventional practice (Miller, R. E., 1981). However, these have not proved to be better than solid urea or ammonium nitrate and are not used in commercial practice.

Other organic sources, such as sewage sludge, or nitrogen fixing plants and species mixtures have been covered in Chapter 4.

Compound fertilisers

Compound fertilisers are not commonly used in forestry due to their high production cost. However, there may be occasions where this additional cost may be justified, e.g. unavailability of normal supplies or hand applications in remote locations. One major exception would be the use of a granular fertiliser containing rock phosphate and muriate of potash to improve aerial distribution. Although more expensive than a mechanical mixture of the two components, it overcomes any problems associated with settling during storage and handling.

If it is planned to use compound fertilisers then the targets for the rates of nutrients applied remain the same (i.e. 150 kg N, 60 kg P and 100 kg K per treated ha). Compound fertilisers are quoted as being '0:20:20', '20:14:14'. These refer to the percentage of N, P_2O_5 and K_2O present in the product. The latter values are converted from P_2O_5 to P by multiplying by 0.44 and from K_2O to K by multiplying by 0.83. To obtain the application rate of compound fertiliser which will be closest to the desired target rates for the individual nutrients the following equation should be used:

$$R = \frac{(150n + 26.4p + 83k) \times 100}{n^2 + 0.1936p^2 + 0.689k^2}$$

where, R = application rate of compound fertiliser (kg per treated ha) and n, p and k refer to the ratios quoted for N, P and K (i.e. n:p:k).
(*Note:* This equation can still be used when only two of the components are present in the compound fertiliser.)

To calculate the quantities (kg per ha) of each nutrient applied:

$N = (R \times n) \div 100$
$P = (R \times p \times 0.44) \div 100$
$K = (R \times k \times 0.83) \div 100$

A worked example is given below:

Compound NPK fertiliser = 20:14:14

$$R = \frac{[(150 \times 20) + (26.4 \times 14) + (83 \times 14)] \times 100}{400 + (0.1936 \times 196) + (0.689 \times 196)}$$

$$= \frac{453\,160}{572.99}$$

= 791 kg per ha (supplying: 158 kg N, 49 kg P and 92 kg K per ha).

Chapter 6
Methods of Application

Most fertiliser application is by helicopter. However, treatment of areas smaller than 20 ha and a sizeable proportion of applications in the establishment phase are carried out by ground based methods (mechanical or manual). All these systems employ a broadcast method of application, avoiding watercourses, as there is no real growth advantage in spot or strip applications with the main fertiliser used – phosphate (Mackenzie, 1972; McIntosh, 1984b). Broadcasting the fertiliser should also encourage rooting development over the whole site which has potential advantages in encouraging full site exploitation and for crop stability (C. P. Quine, personal communication).

Aerial application

Helicopters are normally used and the fertiliser is distributed from a suspended hopper through a powered spinner (Farmer et al., 1985). The hopper usually has a capacity of between 500 and 1000 kg from which 14–20 tonnes of fertiliser can be spread per hour. One of the main problems is the difficulty of achieving even distribution due to the rough terrain and varying windspeeds in many of the application areas. This is exacerbated by the pilot being unable to maintain correct flight lines and start or finish points for the different loads.

The main recent development has been the testing of electronic positioning systems (Hedderwick and Will, 1982; Farmer et al., 1985; Potterton, 1990). This comprises a transmitter/receiver and computer in the helicopter, and two or more transponders (ground based beacons) which are placed on high ground around the treatment area. Radio signals are transmitted from the transponders to the helicopter where they are interpreted by the computer. The computer then determines the position of the helicopter and provides a display to guide the pilot on the correct flight line. In addition to this the computer produces a map of the flight lines which can be used to check the quality of flying. Problems occur with this system particularly when ground conditions cause interference on the signals. In these situations the pilot has to revert to flying by eye.

Provided the navigation system operates correctly, this should allow the pilot to fly regular flight lines over the whole area. However, uniform application is also dependent on the speed of travel and the ideal system would be one which adjusts the hopper aperture in relation to ground speed. (This is being developed in New Zealand.)

The type of fertiliser applied is another important variable. For example, it is not possible to achieve an even distribution with unground rock phosphate (Farmer et al., 1985) (Plate 16) and there are examples of irregular crop caused by 'banded' application (Taylor, 1990a) (Plate 17). On very infertile sites this will result in areas of check and consequent severe loss of timber yield. A granular material is more suitable (Barker, 1979; Eyre, 1982), and although it is more costly this can be offset by having a more uniform crop as a result of improved distribution. The other major advantage is the greater certainty that fertiliser will be contained within the target area and not drift from the site. This is not only wasteful but could result in contamination of watercourses or adjacent sites of conservation value.

In areas where there is a complex array of

site types which require different fertiliser inputs it is possible to target fertiliser application to reduce costs. For example, a particular crop may require an overall top dressing of phosphate and potassium but there may be patches of check induced by nitrogen deficiency. It would not be cost-effective to apply an overall topdressing of nitrogen, but if the patches were not treated they would not be harvestable at the same time as the rest of the crop. In such cases the preferred option would be to carry out a targetted application of nitrogen to the checked areas only. Aerial photography and/or soil maps should be used to estimate and map the areas requiring targetted applications.

Application from the ground

Tractor application is done either before or in conjunction with site preparation prior to planting. The distribution of fertiliser can be very precise but it will take longer than aerial application. There is increased supervisory time and there are greater logistical problems with layout of fertiliser. In many cases technical difficulties associated with achieving an even spread rate and regular swathes have resulted in banding of fertiliser and the results have often not been a great improvement on aerial fertilising.

Manual distribution is limited to areas where the crops are less than 2 m in height and access is still reasonably easy. This method also requires longer periods of supervision than aerial application but, provided that there is correct supervision, it is the most accurate application method. The operation can be greatly facilitated by using helicopters or tractors to deliver fertiliser to a well planned distribution of dump sites. Higher labour costs are offset against the cost of helicopter application and the reduced amount of fertiliser applied to the

Plate 16. *Fertiliser distribution trials showing drift of unground rock phosphate despite ideal conditions.* (R. McIntosh)

Plate 17. Uneven crop growth due to uneven aerial application of fertiliser. (P. M. Tabbush)

target area. This is due to the fact that, unlike aerial application, hand application can avoid placing fertiliser in plough furrows. Therefore, although the rate applied to the treated area remains the same the total amount of fertiliser used over the gross area can be reduced by 20% (estimated area occupied by plough furrows). It is also possible to calculate the application rate for individual trees. For example, at a planting density of 2500 per ha:

Unground rock phosphate = 450 kg per ha × 0.8 ÷ 2500 = 144 g per tree (360 kg per ha)
Urea = 350 kg per ha × 0.8 ÷ 2500 = 112 g per tree (280 kg per ha)

Hand application of fertiliser should be used where it is important to ensure even distribution, e.g. initial application to nursing mixtures on infertile sites, or where it is vital to contain the fertiliser within the treated area, e.g. sensitive water catchments.

When hand applying fertiliser the tree should not be used as a target for handfuls of fertiliser; applying nitrogen, soluble phosphates or potassium in this way will kill small trees. Instead, the appropriate amount of fertiliser should be broadcast evenly around the tree; this will also encourage the roots to spread away from the planting position.

Chapter 7
Economic Appraisal

In British forestry one of the primary objectives is to produce timber as economically as possible. Therefore any major forest operation, such as the application of fertiliser, must provide a net financial benefit. However, rotation lengths of 45–80 years for conifers and 70–120 years for broadleaves make investment appraisal difficult. The long delay between operational expenditures and revenue from harvesting is further complicated by the fact that both expenditure and revenue can occur at different stages in the crop rotation. To allow for this, economic appraisal is based on discounting all expenditures and revenues back to provide present values. Different managerial options can then be compared by calculating benefit:cost ratios (Busby and Grayson, 1981).

There are three possible methods of analysis for appraising investment in fertiliser:
- applying the 'years saved' method;
- comparing recorded height gains with those necessary to justify the costs of fertilising; or
- assuming a permanent change in Yield Class.

The 'years saved' method

Currently, the economic benefits of fertilising are generally appraised on the measured differences in height growth between fertilised and unfertilised crops. This is translated into a reduction in rotation length, or the number of 'years saved' (Everard, 1974). If the increase in discounted revenue associated with this shorter rotation is greater than the discounted cost of the fertiliser application then the operation is considered worthwhile.

There are two main difficulties with this approach:
(a) For how many years after application of fertiliser should the differences in height growth between fertilised and unfertilised crops be assessed?
(b) How do these differences in growth develop during the life of crops and how do they translate into differences in productivity?

This method assumes that observed differences in early height growth do not increase with time. Response to application of fertiliser is regarded as only a transitory boost to growth after which crops return to their 'original' (i.e. unfertilised) growth patterns (Insley et al., 1987; Whiteman, 1988).

Justifying fertiliser costs

An alternative approach is to calculate the improvements in growth necessary to justify the costs of fertilising (i.e. the height gain necessary to make net present value (NPV) = 0) and then to assess the likelihood of these increases in growth being achieved (Insley et al., 1987). The number of years saved on the rotation which are necessary to justify these costs are then calculated from the equation (Whiteman, 1988):

$$1.05^S = DC \div NDR + 1$$

where s = number of years to be saved
DC = discounted costs of fertilising
NDR = net discounted revenue for unfertilised crop (see Whiteman, 1988).

The number of 'years saved' can then be converted into the height gain necessary to justify the fertiliser cost. This height gain can then be judged against local experience of the likely actual response.

Both the above methods are strongly dependent on the Yield Class of the (unfertilised) crops because of the influence this exerts on net discounted revenue. Such methods, while they have the virtue of being based on clear field evidence, disregard the possibility of longer lasting increases in growth as a result of fertilising. However, recent evidence indicates that growth responses to fertiliser do continue to develop with time (Taylor and Worrell, 1991b).

Permanent change in Yield Class

This method assumes that application of fertiliser results in a permanent increase in productivity which can be expressed as a change in Yield Class. On the poorest sites, such as unflushed deep peats, fertiliser is essential for tree growth and will result in a permanent change in Yield Class. On better sites the effects will be smaller and more transitory although in many cases fertiliser applications have been shown to be financially justified (Taylor and Worrell, 1991b).

From the viewpoint of the forest manager this is the most useful method for assessing the benefits of fertiliser application. Unfortunately, there are not yet sufficient data from long-term experiments to enable this approach to be applied in all situations. However, there is currently enough evidence on the response of crops to fertiliser on a range of site types to enable the prediction of where higher levels of productivity can be expected and further appraisal should not be necessary (see Table 3). In the other situations where fertiliser application has been indicated as potentially beneficial an appraisal should be carried out using the method of justifying fertiliser costs.

Chapter 8
Environmental Effects

The main environmental concern associated with application of fertiliser is that enhanced levels of nutrients in run-off could lead to water bodies becoming unsuitable as drinking supplies or could promote algal blooms. Algal blooms are a potential problem in upland waters, where biological activity is usually phosphate-limited (Harriman, 1978), and large increases in phosphate levels could result in a dramatic rise in algal population. This causes a reduction in dissolved oxygen during the night and an increase in pH during the day, both of which can kill fish (Anon., 1988). As yet, however, no instance of serious contamination from forest fertiliser use has been recorded in Britain.

The greatest danger exists where fertilisers are applied at time of planting. At this stage the root systems of the trees have not developed and, where ploughing is used in new planting, the plough furrows are bare of vegetation. The furrows occupy about 20% of the area and fertiliser falling in these is the most liable to wash-off. Although unground rock phosphate is slowly soluble and much of the phosphorus will be retained in the soil or taken up by the trees and the ground vegetation, losses can be quite high from organic soils (Fox and Kamprath, 1971; Malcolm et al., 1977). Therefore care needs to be exercised when applying phosphate in certain water catchment areas, particularly those with closed bodies of water (Harriman, 1978), with a high proportion of peat soils. Although the other main fertiliser, muriate of potash, is readily soluble and potassium leaches from peat soils (Malcolm and Cuttle, 1983) it should not pose any water pollution problems.

Compared with agricultural practice, the risk of phosphate contamination is reduced by other factors such as the infrequency of applications due to the length of the response period, the limitation of fertiliser application to the first 20 years of the rotation and the adequate nutrient supply already present on many forest sites. There are also the standard practices of leaving unfertilised buffer strips around natural watercourses and avoiding application on to frozen or snow-covered ground. In sensitive catchments fertiliser applications should be staggered over a number of years and applied by hand.

Nitrate contamination is unlikely to be a problem in Britain because nitrogen fertiliser is not widely applied (Table 1). When nitrogen is applied the trees are at least 6 years old (McIntosh, 1983b) with well developed roots, and the plough furrows are less exposed and carrying much less water. Nitrogen is also effectively retained on site (Miller et al., 1976a; Morrison and Foster, 1977) particularly the nitrogen in urea which is converted to ammonium and firmly held by the organic matter (Overrein, 1970).

The acidification of watercourses is a serious problem on infertile sites over acid, poorly buffered lithologies (Nisbet, 1990), including areas such as granite outcrops and Silurian and Ordovician sediments in upland Wales (Edmunds and Kinniburgh, 1986). Liming has been advocated as a remedial solution to alleviate this problem, and the most cost-effective methods appear to be direct liming of lakes or of the source areas in catchment headwaters (Nisbet, 1989). Liming of whole catchments is unlikely to be cost-effective and can be damaging to existing plant communities as well as possibly retarding tree growth (McIntosh, 1983b; Derome et al., 1986).

Local water authorities should always be consulted before commencing a fertiliser operation in a water catchment (Anon., 1988).

Chapter 9
Effects on Wood Quality

Although many wood characteristics are under genetic control, silvicultural operations also have their effects. The properties which are most likely to be influenced by fertilisation are density, proportion of latewood, fibre characteristics and the juvenile core (Bevege, 1984).

The density of wood is extremely important, with lower strength and value being associated with less dense timber. Whilst fertiliser application may cause variation in the density of different parts of the stem (McKinnell, 1972; Shepard, 1982), the major concern is the production of irregular growth patterns by badly timed fertiliser applications on nutrient poor sites. However, it is unlikely that the growth benefits from normal, properly-timed applications will reduce wood density below that of untreated stands on fertile sites.

The proportion of latewood in each annual growth ring normally increases with the age of the tree, thus increasing wood strength. This natural process can be delayed by fertilisation (Bamber, 1972), but this is unlikely to cause major reductions in wood quality. Similarly, the shortening of tracheid lengths which has been observed in some species following fertilisation (Bevege, 1984) should not greatly influence wood quality.

Rapid early growth can produce a wide inner core of juvenile wood which has lower strength than mature wood. This could obviously be affected by fertilisation but other silvicultural practices, such as plant spacing, are likely to have a greater influence. Furthermore, this problem is likely to be of greater concern in short-rotation forestry (less than 20 years) rather than in the medium length rotations normally practised in Britain (at least 40 years).

ACKNOWLEDGEMENTS

Many people have contributed to the preparation of this Bulletin. Previous Silviculture (North) project leaders, particularly Bob McIntosh, John Mackenzie, Jim Atterson and Graham Mayhead, established the experiments from which most of my ideas were derived. Numerous research foresters and workers have tended and measured these experiments over the years. Bill Binns, sadly no longer with us, provided advice and encouragement in the author's early struggle to grasp this subject. So did many other colleagues at various research establishments, particularly Michael Carey, Dave Dickson, Dick McCarthy, Douglas Malcolm, Hugh Miller and Mike Proe. Iain White has always been able to redirect errant thoughts on statistics, and produced the equations in Chapter 5. The grouping of soil types followed earlier work by Bob McIntosh. The estimation of lithological influence was developed with Rick Worrell following discussions with Graham Pyatt. Finally, thanks are due to Ian Forrest and David Paterson for reviewing the draft of this Bulletin.

REFERENCES

ANON. (1988). *Forests and water guidelines.* Forestry Commission, Edinburgh.

BALLARD, R. (1981). Urea and ammonium nitrate as nitrogen sources for southern pine plantations. *Southern Journal of Applied Forestry* **5**, 105-108.

BALLARD, R. (1984). Fertilisation of plantations. In *Nutrition of plantation forests*, eds G. D. Bowen and E. K. S. Nambiar, 327-360. Academic Press, London.

BAMBER, R. K. (1972). Some studies of the effects of the fertilisers on the wood properties of *Pinus* species. In *The Australian forest-tree nutrition conference 1971*, ed. R. Boardman, 366-379.

BARKER, P. R. (1979). Factors affecting aerial distribution of fertilizers to forests. In *Forest fertilization conference*, eds S. P. Gessel, R. M. Kenady and W. A. Atkinson, 196-204. University of Washington, Seattle.

BASTION, R. K. (1986). Overview on sludge utilisation. In *The forest alternative for the treatment and utilization of municipal and industrial waste*, eds D. W. Cole, C. L. Henry and W. L. Nutter, 7-25. University of Washington Press, Seattle.

BAULE, H. and FRICKER, C. (1970). *The fertiliser treatments of forest trees.* Translated by C. I. Whittles; 259 pp. BLV, Munich.

BAYES, C. D., TAYLOR, C. M. A. and MOFFAT, A. J. (1989). Sewage sludge utilisation in forestry: the UK research programme. In *Alternative uses of sewage sludge*, University of York, 5-7 Sept. 1989, ed. J. E. Hall. Water Research Centre, Medmenham, UK.

BAYES, C. D. and TAYLOR, C. M. A. (1989). *The use of sewage sludge in forestry – a draft code of practice.* Water Research Centre Report UM 1019. Water Research Centre Medmenham, UK.

BEVEGE, D. I. (1984). Wood yield and quality in relation to tree nutrition. In *Nutrition of plantation forests*, eds G. D. Bowen and E. K. S. Nambiar, 293-326. Academic Press, London.

BENZIAN, B. and WARREN, R. G. (1956). Copper deficiency in Sitka spruce seedlings. *Nature* **178**, 864-865.

BENZIAN, B. (1965). *Experiments on nutrition problems in forest nurseries.* Forestry Commission Bulletin 37. HMSO, London.

BINNS, W. O. (1975). *Fertilisers in the forest: a guide to materials.* Forestry Commission Leaflet 63. HMSO, London.

BINNS, W. O., MAYHEAD, G. J. and MACKENZIE, J. M. (1980). *Nutrient deficiencies of conifers in British forests.* Forestry Commission Leaflet 76. HMSO, London.

BINNS, W. O. and FOURT, D. F. (1981). Surface workings and trees. In *Research for practical arboriculture*. Proceedings of the Forestry Commission/Arboricultural Association Seminar, Preston, Feb. 1980, 60-75. Foresty Commission Occasional Paper 10. Forestry Commission, Edinburgh.

BINNS, W. O., INSLEY, H. and GARDINER, J. B. H. (1983). *Nutrition of broadleaved amenity trees. I. Foliar sampling and analysis for determining nutrient status.* Arboriculture Research Note 50/83/SSS. DOE Arboricultural Advisory and Information Service, Forestry Commission, Edinburgh.

BOULD, C., HEWITT, E. J. and NEEDHAM, P. (1983). *Diagnosis of mineral disorders in plants.* Vol. 1; 170 pp. HMSO, London.

BUSBY, R. J. N. and GRAYSON, A. J. (1981). *Investment appraisal in forestry.* Forestry Commission Booklet 47. HMSO, London.

CAREY, M. L. (1980). Whole tree harvesting in Sitka spruce. Possibilities and implications. *Irish Forestry* **37**, 48-63.

CAREY, M. L., McCARTHY, R. G. and MILLER, H. G. (1988). More on nursing mixtures. *Irish Forestry* **45**, 7-20.

CARLYLE, J. C. and MALCOLM, D. C. (1986). Nitrogen availability beneath pure spruce and mixed larch + spruce stands growing on a deep peat. *Plant and Soil* **93**, 115-122.

COUTTS, M. P., LOW, A. J., PYATT, D. G., BINNS, W. O. and CARTER, C. I. (1985). Reduced growth and bent top of Sitka spruce. *Forestry Commission Report on Forest Research 1985*, 32. HMSO, London.

DANGERFIELD, J. and BRIX, H. (1981). Comparative effects of ammonium nitrate and urea fertilisers on tree growth and soil processes. In *Forest fertilisation conference*, eds S. P. Gessel, R. M. Kenady and W. A. Atkinson, 133-139. University of Washington, Seattle.

DAVEY, C. B. and WOLLUM, A. G. (1984). Nitrogen fixation systems in forest plantations. In *Nutrition of plantation forests*, eds G. D. Bowen and E. K. S. Nambiar, 361-377. Academic Press, London.

DEROME, J., KUKKOLA, M. and MALKONEN, E. (1986). *Forest liming on mineral soils*. National Swedish Environment Protection Board Report, 3084; 107 pp. Sweden.

DICKSON, D. A. (1977). Nutrition of Sitka spruce on peat – problems and speculations. *Irish Forestry* **34**, 31-39.

DIGHTON, J. and HARRISON, A. F. (1983). Phosphorus nutrition of lodgepole pine and Sitka spruce stands as indicated by a root bioassay. *Forestry* **56**, 33-43.

DUTCH, J. C., TAYLOR, C. M. A. and WORRELL, R. (1990). *Potassium fertiliser – effects of different rates, types and times of application on height growth of Sitka spruce on deep peat*. Forestry Commission Research Information Note 188. Forestry Commission, Edinburgh.

EDMUNDS, W. M. and KINNIBURGH, D. G. (1986). The susceptibility of UK groundwaters to acidic deposition. *Journal of the Geological Society of London* **143**, 707-720.

EDWARDS, M. V. (1958). Use of triple superphosphate for forest manuring. *Forestry Commission Report on Forest Research 1958*, 117-130. HMSO, London.

EVERARD, J. E. (1973). Foliar analysis sampling methods, interpretation and application of the results. *Quarterly Journal of Forestry* **67**, 51-66.

EVERARD, J. E. (1974). *Fertilisers in the establishment of conifers in Wales and southern England*. Forestry Commission Booklet 41. HMSO, London.

EVANS, J. (1986a). Nutrition experiments in broadleaved stands: I. Pole-stage ash and oak. *Quarterly Journal of Forestry* **80**, 85-94.

EVANS, J. (1986b). Nutrition experiments in broadleaved stands: II. Sweet chestnut and stored oak coppice. *Quarterly Journal of Forestry* **80**, 95-103.

EYRE, J. R. (1982). *Helicopter topdressing of forests*. Unpublished David Henry Scholarship Report of study tour to Scandinavia, Britain and the USA; 61 pp. Forest Service, Auckland, New Zealand.

FARMER, R. A., ALEXANDER, A. and ACTON, M. (1985). Aerial fertilising: monitoring spread is vital operation. *Forestry and British Timber* **14**, 15-17.

FISHER, R. F. and PRITCHETT, W. L. (1982). Slash pine growth response to different nitrogen fertilisers. *Soil Science Society of America Journal* **46**, 133-136.

FOX, R. L. and KAMPRATH, E. J. (1971). Adsorption and leaching of P in organic soils and high organic matter sand. *Soil Science Society of America Proceedings* **35**, 154-156.

FRASER, G. K. (1933). *Studies of certain Scottish moorlands in relation to tree growth*. Forestry Commission Bulletin 15. HMSO, London.

GESSEL, S. P. and TURNER, J. (1974). Litter production by red alder in western Washington. *Forest Science* **20**, 325-330.

GRIFFIN, E., CAREY, M. L. and McCARTHY, R. G. (1984). Treatment of checked Sitka spruce crops in Republic of Ireland. In *Aspects of Applied Biology* **5**, *Weed control and vegetation management in forests and amenity areas*, ed. P. M. Tabbush, 211-222. Wellesbourne, Warwick, UK.

HARRIMAN, R. (1978). Nutrient leaching from fertilised forest watersheds in Scotland. *Journal of Applied Ecology* **15**, 933-942.

HARRINGTON, C. A. and MILLER, R. E. (1979). *Response of a 110-year-old Douglas fir stand to urea and ammonium nitrate fertilisation*. USDA Forest Service Research Note PNW-336; 7pp.

HEDDERWICK, G. W. and WILL, G. M. (1982). *Advances in the aerial application of fertiliser in New Zealand forests: the use of an electronic guidance system and dust-free fertiliser*. Forest Research Institute Bulletin 34. Roturua, New Zealand.

INSLEY, H., HARPER, W. G. C. and WHITEMAN, A. (1987). *Investment appraisal handbook*. Forestry Commission, Edinburgh.

JOBLING, J. and STEVENS, F. R. W. (1980). *Establishment of trees on regraded colliery spoil heaps*. Forestry Commission Occasional Paper 7. Forestry Commission, Edinburgh.

JONES, H. E., HARRISON, A. F. and DIGHTON, J. (1987). A ^{86}Rb bioassay to determine the potassium status of trees. *New Phytologist* **107**, 695-708.

LINES, R. (1987). *Choice of seed origins for the main forest species in Britain*. Forestry Commission Bulletin 66. HMSO, London.

LINES, R. and BROWN, I. (1982). Broadleaves for the uplands. In *Broadleaves in Britain – future management and research*, eds D. C. Malcolm, J. Evans and P. N. Edwards. Institute of Chartered Foresters, Edinburgh.

McKINNELL, F. H. (1972). Fertiliser and wood density of *Radiata* pine. In *Australian forest-tree nutrition conference, 1971*, ed. R. Boardman, 381-383.

McINTOSH, R. (1981). Fertiliser treatment of Sitka spruce in the establishment phase in upland Britain. *Scottish Forestry* **35**, 3-13.

McINTOSH, R. (1983a). *The effect of fertiliser and herbicide applications on Sitka spruce on mineral soils with a dominant grass/herb vegetation*. Forestry Commission Research Information Note 76/83/SILN. Forestry Commission, Edinburgh.

McINTOSH, R. (1983b). Nitrogen deficiency in establishment phase Sitka spruce in upland Britain. *Scottish Forestry* **37**, 185-193.

McINTOSH, R. (1984a). *Fertiliser experiments in established conifer stands*. Forestry Commission Forest Record 127. HMSO, London.

McINTOSH, R. (1984b). *Phosphate fertilisers in upland forestry – types, application rates and placement methods*. Forestry Commission Research Information Note 89/84/SILN. Forestry Commission, Edinburgh.

MACKENZIE, J. M. (1972). Early effects of different types, rates and methods of application of phosphate rock on peatland. In *The Proceedings of the 4th international peat congress*, Vol III, 531-546.

MACKENZIE, J. M. (1974). Fertiliser/herbicide trials on Sitka spruce in east Scotland. *Scottish Forestry* **28**, 211-221.

McNEILL, J. D. and MOFFAT, A. J. (1989). Reclamation of upland sites. *Forestry Commission Report on Forest Research 1989*, 25. HMSO, London.

MADGWICK, H. A. I., JACKSON, D. S. and KNIGHT, P. J. (1977). Above-ground dry matter, energy and nutrient contents of trees in an age series of *Pinus radiata* plantations. *New Zealand Journal of Forestry Science* **7**, 445-468.

MALCOLM, D. C. (1972). The effect of repeated urea applications on some properties of drained peat. *Proceedings of the international peat congress, Finland* **3**, 451-460.

MALCOLM, D. C. (1975). The influence of heather on silvicultural practice – an appraisal. *Scottish Forestry* **29**, 14-24.

MALCOLM, D. C., BRADBURY, I. K. and FREEZAILLAH, B. C. Y. (1977). The loss of nutrients from fertilised peat in an artificial system in the field. *Forest Ecology and Management* **1**, 109-118.

MALCOLM, D. C. and CUTTLE, S. P. (1983). The application of fertilisers to drained peat. I. Nutrient losses in drainage. *Forestry* **56**, 155-174.

MALCOLM, D. C. and TITUS, B. D. (1983). Decomposing litter as a source of nutrients for second rotation stands of Sitka spruce established on peaty gley soils. In *Forest site and continuous productivity*, eds R. Ballard and S. P. Gessel, 138-145. USDA Forest Service General Technical Report PNW-163.

MEAD, D. J., BALLARD, R. and MACKENZIE, M. (1975). Trials with sulfur-coated urea and other nitrogenous fertilisers on *Pinus radiata* in New Zealand. *Soil Science Society of America Proceedings* **39**, 978-980.

MILLER, H. G. (1981). Forest fertilisation: some guiding concepts. *Forestry* **54**, 157-167.

MILLER, H. G. (1984). Dynamics of nutrient cycling in plantation ecosystems. In *Nutrition of plantation forests*, eds G. D. Bowen and E. K. S. Nambiar, 53-78. Academic Press, London.

MILLER, H. G., COOPER, J. M. and MILLER,

J. D. (1976a). Effect of nitrogen supply on nutrients in litter fall and crown leaching in a stand of Corsican pine. *Journal of Applied Ecology* **13**, 233-248.

MILLER, H. G., MILLER, J. D. and PAULINE, O. J. L. (1976b). Effect of nitrogen supply on nutrient uptake in Corsican pine. *Journal of Applied Ecology* **13**, 955-966.

MILLER, R. E. (1981). Response of Douglas fir to foliar fertilisation. In *Forest fertilisation conference*, eds S. P. Gessel, R. M. Kenady and W. A. Atkinson, 62-68. University of Washington, Seattle.

MOLLER, G. (1974). The choice of fertilisers and the timing of application. *Skogen* **61**, 80-89.

MORRISON, I. K. and FOSTER, N. W. (1977). Fate of urea fertiliser added to a boreal forest *Pinus banksiana* Lamb. stand. *Soil Science Society of America Journal* **41**, 441-448.

NICHOLS, C. G. (1989). US forestry uses of municipal sewage sludge. In *Alternative uses of sewage sludge*, University of York, 5-7 Sept. 1989, ed. J. E. Hall. Water Research Centre, Medmenham, UK.

NIMMO, M. and WEATHERELL, J. (1961). Experiences with leguminous nurses in forestry. *Forestry Commission Report on Forest Research 1961*, 126-147. HMSO, London.

NISBET, T. R. (1989). *Liming to alleviate surface water acidity*. Forestry Commission Research Information Note 148. Forestry Commission, Edinburgh.

NISBET, T. R. (1990). *Forests and surface water acidification*. Forestry Commission Bulletin 86. HMSO, London.

O'CARROLL, N. (1978). The nursing of Sitka spruce. 1. Japanese larch. *Irish Forestry* **35**, 60-65.

OTCHERE-BOATENG, J. (1981). Reaction of nitrogen fertilisers in forest soils. In *Forest fertilisation conference*, 1981, eds S. P. Gessel, R. M. Kenady and W. A. Atkinson, 37-42. University of Washington, Seattle.

OVERREIN, L. N. (1970). Tracer studies on nitrogen immobilisation/mineralisation relationships in forest raw humus. *Plant and Soil* **32**, 478-500.

OVINGTON, J. D. (1957). Dry-matter production by *Pinus sylvestris* L. *Annals of Botany* **82**, 287-314.

POTTERTON, E. (1990). Electronic track guidance comes of age in British forestry. *Forestry and British Timber* **19** (2), 33, 35.

PYATT, D. G. (1982). *Soil classification*. Forestry Commission Research Information Note 68/82/SSN. Forestry Commission, Edinburgh.

RADWAN, M. A., DEBELL, D. S., WEBSTER, S. R. and GESSEL, S. P. (1984). Different nitrogen sources for fertilising western hemlock in western Washington. *Canadian Journal of Forest Research* **14**, 155-162.

SHEPARD, R. K. (1982). Fertilisation effects on specific gravity and diameter growth of red spruce. *Wood Science* **14**, 138-144.

STIRLING-MAXWELL, J. (1925). On the use of manures in peat planting. *Transactions of the Royal Scottish Arboricultural Society* **XXXIX**, 103-109.

TAMM, C. O., HOLMEN, H., POPOVIC, B. and WIKLANDER, G. (1974). Leaching of plant nutrients from soils as a consequence of forestry operations. *Ambio* **III(6)**, 211-221.

TAYLOR, C. M. A. (1985). The return of nursing mixtures. *Forestry and British Timber* **14**, 18-19.

TAYLOR, C. M. A. (1986). *Forest fertilisation in Great Britain*. Fertiliser Society Proceedings 251; 23pp. Fertiliser Society, London.

TAYLOR, C. M. A. (1987). Effects of nitrogen fertiliser at different rates and times of application on the growth of Sitka spruce in upland Britain. *Forestry* **60**, 87-99.

TAYLOR, C. M. A. (1989). Copper deficiency. *Forestry Commission Report on Forest Research 1989*, 18. HMSO, London.

TAYLOR, C. M. A. (1990a). Survey of fertiliser prescriptions in Scotland. *Scottish Forestry* **44**, 3-9.

TAYLOR, C. M. A. (1990b). *The nutrition of Sitka spruce on upland restock sites*. Forestry Commission Research Information Note 164. Forestry Commission, Edinburgh.

TAYLOR, C. M. A. and MOFFAT, A. J. (1989). The potential for utilising sewage sludge in forestry in Great Britain. In *Alternative uses of sewage sludge*, University of York, 5-7 Sept. 1989, ed. J.E. Hall. Water Research Centre, Medmenham, UK.

TAYLOR, C. M. A. and TABBUSH, P. M. (1990). *Nitrogen deficiency in Sitka spruce plantations*. Forestry Commission Bulletin 89. HMSO, London.

TAYLOR, C. M. A. and WORRELL, R. (1991a). Influence of site factors on the response of Sitka spruce to fertiliser at planting in upland Britain. *Forestry* **64** (1), 13-27.

TAYLOR, C. M. A. and WORRELL, R. (1991b). *Medium-term responses to fertiliser – implications for economic viability*. Forestry Commission Occasional Paper (in press).

TITUS, B. D. and MALCOLM, D. C. (1987). The effect of fertilization on the decomposition in clearfelled spruce stands. *Plant and Soil* **100**, 297-322.

WEATHERELL, J. (1954). Checking of forest trees by heather. *Forestry* **26**, 37-40.

WEETMAN, G. F. and ALGAR, D. (1974). Jack pine nitrogen fertilisation and nutrition studies: 3-year results. *Canadian Journal of Forest Research* **4**, 381-398.

WHITEMAN, A. (1988). Appraising fertiliser application – the case of the Moine schists. *Scottish Forestry* **42**, 255-275.

WILLIAMSON, D. R. and LANE, P. B. (1989). *The use of herbicides in the forest*. Forestry Commission Field Book 8. HMSO, London.

WRIGHT, T. W. and WILL, G. M. (1958). The nutrient content of Scots and Corsican pines growing on sand dunes. *Forestry* **31**, 13-25.

ZEHETMAYR, J.W. L. (1954). *Experiments in tree planting on peat*. Forestry Commission Bulletin 22. HMSO, London.

ZEHETMAYR, J. W. L. (1960). *Afforestation of upland heaths*. Forestry Commission Bulletin 32. HMSO, London.

ZULU, L. C. (1989). *Copper deficiency in malformed spruce*. Unpublished Hons. Thesis, Edinburgh University; 46pp.

Appendix I

Foliar Sampling Procedure

Introduction

1. The foliage sampling procedure described below is designed primarily for Sitka spruce, lodgepole and Scots pine in upland Britain, but also includes reference to larch and broadleaves.
2. It must be stressed that foliage analysis levels are not an infallible guide to achieving financially optimal inputs of fertiliser. Soil type, current tree growth rate and appearance, past fertiliser treatment, proximity to canopy closure, species and local experience are all important factors to be considered in deciding whether to fertilise. In many cases consideration of these other factors may be adequate and the cost and delay inherent in sampling thus avoided.
3. At present foliar sampling should be restricted to stands up to 3 – 4 m mean height. Where sampling is being considered for taller stands the Forestry Commission Research Division should be consulted.

Identification of areas to sample

4. Although compartments can be sampled individually, foliage analysis results will be more useful if they are derived from a sampling programme designed to examine a particular problem. Sampling might for example be aimed to assess or monitor the nutritional status of a certain age class of Scots pine on a particular soil type.
5. Having identified the problem and the crop to be sampled it should then be subdivided into areas of good and poor growth. Foliar samples should be taken from both these areas as relative nutrient values assist in the interpretation of the results of analysis. Ideally, several sets of composite samples should be collected from each category to improve accuracy of results.

Timing of sampling

6. Samples may be collected at any time of day from the first week in October to the end of the second week in November. Deciduous conifers and broadleaves should be sampled in late July or August after shoot growth is complete and before needles or leaves begin to change colour.
7. It is desirable to avoid sampling after periods of prolonged rain because nutrients, particularly potassium, may leach from the needles or leaves. However this may not always be possible and collection of samples will depend on local weather conditions and previous experience.

Selection of trees to sample

8. Five typical dominant and co-dominant trees should be sampled to form a composite set of samples within each category.

Age of foliage

9. Only shoots of the current year should be taken; needles or leaves should not be stripped off the shoot.

Position of sample shoot

10. On conifers sample shoots should be taken from the first whorl below the leader, excluding any lammas growth; one shoot is taken per tree. Undamaged, fully expanded leaves should be collected from the outside of the crown of broadleaved trees. Shoots or leaves must be fully exposed to sunlight as shading affects the nutrient content.

Size of sample

11. Five sample shoots from conifers is sufficient for analysis, irrespective of the growth rate of the trees sampled. Where shoots are too long to fit in the polythene bag they should be carefully cut in two.
12. About 100 cm^2 of leaves, excluding the petioles, should be collected from broadleaved trees.

Appendix II

Forestry Commission Foliar Analysis Service

Introduction

1. The Forestry Commission Research Division provides a service for foliar analysis and interpretation of results. It will be necessary to check periodically that the instructions detailed below are still appropriate and to obtain the necessary forms.

Packing, labelling and dispatch of samples

2. After samples have been collected (see Appendix I) the shoots should be shaken if wet and then packed, even if moist, into 250 mm x 350 mm polythene bags. A label *must* be included detailing the name of the forest area and the compartment number sampled, the species and planting year. It should also include whether the samples are from good or poor areas. Sets of perforated, ready-printed labels can be obtained from Alice Holt (see below for address).
3. The bag of shoots from 5 trees is rolled up with the label near the top and readable from the outside; the roll should be secured with a moderately tight elastic band. The polythene bags should be neatly but loosely packed into a box.
4. Before dispatch a list of the samples submitted *must* be included, which confirms the information contained on the labels, plus the name and address of the sender, date of dispatch and analysis required (normally NPK).
5. All foliage samples should be immediately sent by first class post to:

    Site Studies (S)
    Forestry Commission Research Station
    Alice Holt Lodge, Wrecclesham
    Farnham
    Surrey
    GU10 4LH

    (Tel: 0420 22255   Fax: 0420 23653)

    The outside of the parcel should be marked "Plant material – urgent".

Site record forms

6. Foliage analysis results on their own may not enable the production of a correct prescription and it is essential to complete a record form (Appendix III) for each site sampled.
7. Site record forms should be dispatched with the foliage samples to Alice Holt.

Results

8. Results and prescriptions for treatment will normally be returned within 6 weeks of samples being received at Alice Holt.

Other problems

9. Analysis for elements other than nitrogen, phosphorus, potassium, calcium and magnesium may require a different technique and Research Division should be consulted. The procedures described in this paper are not suitable for nursery (consult Silviculturist i/c Nursery Project at either the Northern Research Station or Alice Holt) or amenity trees (refer to Arboriculture Research Note 50/83/SSS or consult Arboricultural Advisory and Information Service at Alice Holt).

Appendix III

Forestry Commission Site Record Form for use with Shoot Samples for Foliar Analysis

1. Sender _____   2. Date of dispatch _____

3. Forest District/ _____   4. Collection date _____
   Company/Estate

5. Compartment Nos. _____   6. OS Grid Ref _____

7. Species _____   8. P Year _____

9. Elevation _____   10. Lithology _____
   (OS Geological Map of the UK)

	Areas of good growth	Areas of poor growth
11. Soil type (FC RIN 68/82/SSN)	_____	_____
12. Dominant ground vegetation	_____	_____
13. Crop description		
a. Percentage of crop	_____	_____
b. Stage of development: ☐ pre-thicket ☐ thicket ☐ irregular closure ☐ pole-stage		☐ ☐ ☐ ☐
c. Mean height (m)	_____	_____
d. Needle size: ☐ normal ☐ short		☐ ☐

e. Needle colour: ☐ normal ☐
☐ most needles pale green ☐
☐ most needles yellow ☐
☐ new needles yellow ☐
☐ old needles yellow ☐
☐ old needles yellow and ☐
new needles pale green

f. Needle retention: ☐ several years ☐
☐ only 1 or 2 years ☐

g. Length of leading shoot (cm) _____ Current growth _____

_____ last year's growth _____

14. Treatment at planting

a. Ground preparation: ☐ None
☐ Screefing
☐ Ripping
☐ Mounding
☐ Ploughing

b. Fertilisers: *Product* *Rate*

Phosphate

Potassium

Nitrogen

Other

15. Later treatment   a. Fertilisers: Product   Rate   Year

                                      Phosphate

                                      Potassium

                                      Nitrogen

                                      Other

                         b. Herbicides:

16. Any other relevant information
    e.g. species in mixture, LP provenance, new planting or restocking

Signed ——————————————————   Address ——————————————————

Date ——————————————————

Notes: 1. Foliar samples and the site record form should be sent to Alice Holt (first class post to: Site Studies (S), Alice Holt Lodge, Wrecclesham, Farnham, Surrey, GU10 4LH).
        2. Areas to be sampled should be subdivided into areas of good or poor growth and a minimum of 5 shoots should be sampled within each category. Each set of 5 shoots should be placed in one bag to form a single sample for analysis, together with a label clearly identifying which area and category the sample came from.